Nutritional Epigenetics

CURRO CLAVERO VALDIVIELSO

NUTRITIONAL EPIGENETICS

*The beautiful melody of your genes
in conversation with your environment*

A PRACTICAL GUIDE TO
A LONGER, HEALTHIER LIFE

Foreword by Carlos López-Otín

Title:
   Nutritional Epigenetics
   © Francisco Clavero Valdivielso, 2025
      info@curroclavero.com
      www.curroclavero.com/en

From the foreword:
   © Carlos López-Otín, 2025

ISBN: 978-84-09-73458-0

Design and layout:
   Fran Sánchez Mazo

Cover illustration:
   Marta Blanes Benito

All rights reserved in English. No part of this publication may be reproduced, stored in a retrieval system, or transmitted in any form or by any means without the prior written permission of the copyright holders.

To my family, for all those times when I was with them,
but my mind was here—and for all those times
I was here, but not with them.

To my wife Alma, who saw in me what I had yet to see in myself,
and had the patience to show it to me without ever saying it aloud.

To my children, Jimena and Raúl, who guide me through
the most complex journey of all: being a father.

SPECIAL THANKS

To Carlos López-Otín, my scientific role model, whose ethics, generosity, and humanity surpass even his excellence as a researcher. Thanks to his selfless support, my manuscripts improved immensely. The best parts of this book are thanks to him; the rest is on me.

To my dear friend Clara Pedrejón, whose curiosity and ongoing drive to improve her patients' lives led us to talk about non-pharmacological approaches to autism—conversations that eventually brought us to methylation and how certain nutrients influence epigenetics, a topic that had captivated me for years. That's how the idea for this book was born.

TABLE OF CONTENTS

FOREWORD. *The Epigenetic Grammar of Nutrition*, Carlos López-Otín   15
INTRODUCTION   19

THEORY

CHAPTER 1. **What Is Epigenetics and How Does It Affect Life**   25
    Genetics Vs. Epigenetics: Which Is More Important?   26
    Core Epigenetic Processes   28
    Epigenetic Alterations: Impact on Health and Aging   34

CHAPTER 2. **The Role of SAMe in Epigenetics**   39
    What Is Methylation And The Role Of Same As A Methyl Group Donor   39
    SAMe and DNA Methylation   40
    SAMe and Aging   42

CHAPTER 3. **Other Functions of SAMe and Methylation**   47
    Creatine Synthesis   47
    Phosphatidylcholine Synthesis   48
    Regulation of Circadian Rhythms   49
    Breaking Down Excess Catecholamines and Estrogens   50
    Breaking Down Excess Histamine   51
    Breaking Down Excess Niacinamide (Vitamin B3)   52
    Neurotransmitter Synthesis (Serotonin, Dopamine, and Norepinephrine)   53
    Synthesis of the Hormone and Neurotransmitter Adrenaline   54
    Supporting The Differentiation And Proliferation of T Lymphocytes
        in the Immune System   54

Synthesis of Polyamines: Spermidine and Spermine — 56
Synthesis of Antioxidants: Glutathione and Taurine — 59
Eye Health — 60
Conclusion — 61

CHAPTER 4. **Understanding the Methionine-Homocysteine Cycle** — 65
The Methionine Cycle: How SAMe is Produced — 69
The Homocysteine Cycle: What Happens After Methylation — 70

CHAPTER 5. **Diseases Linked to Global Hypomethylation** — 75
Cardiovascular Diseases — 76
Cancer — 80
Neurodegenerative Diseases — 83
Type 2 Diabetes Mellitus — 98
Depression — 100
Autism — 105
Addictions — 112
Autoimmune Diseases — 114
Allergies — 115
Migraines — 116
Fibromyalgia — 120
Attention Deficit Hyperactivity Disorder (ADHD) — 120
Aging — 121

CHAPTER 6. **Diseases Linked to Global Hypermethylation** — 127
Schizophrenia and Bipolar Disorder — 127
Epilepsy — 130
Acute Myeloid Leukemia (AML) — 132

PRACTICE

CHAPTER 7. **How to Know If You Have a State of Global Hypomethylation or Hypermethylation** — 139
SAMe/SAH Ratio — 140
Methionine/Homocysteine Ratio — 140

CHAPTER 8. How to Increase Methylation in a State of Global Hypomethylation   143

    Supplemental SAMe   143
    Creatine   144
    Phosphatidylcholine   146
    Dietary Choline   146
    Reduce or Avoid Alcohol   147
    Do Not Smoke   147
    Exercise   148
    Sufficient and Quality Sleep   149
    Reduce Endocrine Disruptors: Phthalates and Bisphenol A   149

CHAPTER 9. How to Reduce Methylation in a State of Global Hypermethylation   155

    Glycine   155
    Vitamin B3   157
    Ketogenic Diet   159
    Chlorogenic Acid   160

CHAPTER 10. How to Lower Homocysteine Levels   163

    Vitamins B9 and B12   165
    Trimethylglycine   166
    Vitamin B6   167
    Vitamin B2   169

CHAPTER 11. How to Inhibit Gene Silencing (Deacetylation) of Tumor Suppressor Genes and Others   173

    Alpha-Ketoglutarate   174
    Curcumin   174
    Epigallocatechin-3-Gallate (EGCG)   175
    Allicin   176
    Kaempferol   176
    Proanthocyanidins   177
    Genistein   177
    Resveratrol   180
    Butyrate   180

Chlorogenic and Caffeic Acids	181
Selenium	182
Quercetin	183
Sulforaphane	183
Vitamin D	184
Omega-3 Fatty Acids	185
EPILOGUE	191
APPENDICES	195
Microbiota and Methylation	197
How SAMe and Methylation Affect Physical Performance	201
REFERENCES	225
LIST OF FIGURES	227
ABBREVIATIONS, ACRONYMS, AND INITIALISMS	229

FOREWORD
The Epigenetic Grammar of Nutrition

I have never met Curro Clavero. He has never been my student or my disciple. He hasn't attended my lectures, nor has he lingered at the back of a hall waiting for the right moment to step up to the podium and discreetly ask a question that had stayed with him after my talk. And yet, a few years ago—two or three, perhaps—he wrote to me, sharing his interest in advancing his understanding of the keys to health and the role that nutrition plays in shaping it. From that very first message, Curro sparked a curious sense of empathy in me. His words conveyed sincerity and ease. He made no attempt to hide that his formal academic background in these matters was as a dietitian, but what stood out was his genuine desire to learn beyond that. Time has only confirmed it.

Curro Clavero is a remarkable example of a self-taught scholar in a field as complex as the real impact of nutrition on health and disease. Through immense personal effort, he read extensively and thought deeply. His dedication gave rise to a first book, *Alimentación evolutiva*, which I read with interest and curiosity. Its pages were marked by restraint, free from the grandiosity, exaggeration, or misinformation that floods so many of today's nutrition books on bookstore shelves. Moreover, Curro sought to go further—his book made a real effort to explain the molecular foundations behind the nutritional recommendations he offered readers.

These same principles are what the author has sought to apply in the conception and writing of this new book, *Nutritional Epigenetics*, the pages of which you are about to explore. The task Curro Clavero took on was by no means simple. Epigenetics is a rapidly growing field, and its fundamental concepts are in constant flux—continually revised and redefined. This inherent complexity touches not only on the molecular and cellular aspects of epigenetics but also on how epigenetic changes affect human health.

Curro granted me the privilege of reading the first manuscript of this book. I was struck by the sheer volume of information it contained—overwhelming in its scope—but also concerned, sensing that the book could feel scattered or difficult to navigate for readers unfamiliar with the intricacies of epigenetics. After several exchanges in which I offered suggestions on areas I felt needed reconsideration, followed by a long phone conversation, Curro succeeded in shaping a work that treads carefully through the turbulent waters of epigenetics and the vast *terra incognita* of its many connections to nutrition.

In one of my own books, I described the epigenome as "the grammar or orthography of the genome"—a system in which epigenetic changes act like accents, commas, umlauts, or periods that give grammatical meaning to the genetic message, shaping the dialogue between the genome we inherit from our parents and the environment in which our lives unfold. The epigenome is tasked with organizing the vast information stored in the genome so that, at any given time, only the necessary parts are expressed to keep us tethered to the wheel of survival.

Epigenetic alterations happen continuously—millions of them every moment of life. It's no surprise, then, that depending on our diet, body temperature, physical activity, emotions, or the diseases we may suffer from, these changes are constantly occurring, determining whether genes are switched on or off, whether they speak or remain silent.

Curro Clavero has chosen to focus this book on the epigenetic significance of one of those factors: nutrition—that marvelous elixir of health when its principles are wisely followed to avoid excess, deficiency, or insufficiency.

This book offers a thorough overview of the data currently available in this field, while also clearly acknowledging some of the limitations that still challenge the translation of epigenetic concepts into improved human health. Chief among them are the difficulty in establishing causality from what are often merely correlational findings, and the complexity of achieving precision in the pharmacological—or nutritional—modulation of epigenetic alterations that may underlie functional imbalances in the body.

For all these reasons, I encourage readers to approach *Nutritional Epigenetics* with both focus and curiosity, and to draw from its pages new insights that remind us how malnutrition can shrink our dreams of health—and that our commitment to nutritional education must be radical, rigorous, and unwavering.

Carlos López-Otín*

---

* Carlos López-Otín is Professor of Biochemistry and Molecular Biology, and a member of both the European Academy and the Royal Academy of Sciences of Spain. A world-renowned authority on human genome research and aging, he is the lead author of *The Hallmarks of Aging*, the most influential and widely cited scientific publication in the field of longevity.

INTRODUCTION

Imagine for a moment that your DNA is like an ancient scroll written over millions of years of evolution—a masterpiece that holds the code of your existence. But what if I told you that the real magic doesn't lie only in the words inscribed on that ancestral document, but in the invisible marks that determine which chapters are read and which remain hidden?

This is where epigenetics comes in—a new science that is revolutionizing our understanding of health, disease, and the aging process. Far from being a destiny carved in stone, epigenetics shows us that we have the power to influence how our genes are expressed. And nutrition is the tool that provides the molecules we need to rewrite our story.

At the heart of this molecular dance is methylation, one of the two core epigenetic processes. It acts like a switch, turning genes and regions of DNA on or off depending on the signals it receives from the environment. The silent, yet essential, protagonist of this regulation is S-adenosylmethionine (SAMe), the supplier of the epigenetic "ink" that marks the most relevant genetic instructions at every moment of life.

Health and well-being are, in most cases, not just the product of chance. They are the result of a delicate biological balance taking place within every cell of the body. Through this book, I want to guide you on a journey through the pages of your own book of life and show you how methylation and SAMe availability play a crucial role in:

- Protecting your DNA: by preventing replication errors and preserving its integrity to slow systemic deterioration.
- Optimizing your energy metabolism: by regulating cellular energy production and use.

- Defending against oxidative stress: by boosting the body's natural antioxidants to combat damage.
- Modulating your immune system: by balancing inflammatory responses and enhancing defense mechanisms.
- Promoting longevity: by sustaining cellular vitality, preventing disease, and slowing aging.

You'll discover how the natural decline in SAMe—a process tied to aging—can impair these vital functions and increase the risk of developing a wide range of conditions, from cardiovascular problems to cancer and neurodegenerative or psychiatric diseases.

We'll unravel the secrets of the methionine-homocysteine cycle, a complex metabolic process that regulates the production and recycling of SAMe, and you'll come to understand how the balance of certain nutrients is essential for it to work with the precision of a Swiss watch.

But most importantly, I'll show you that you're no longer at the mercy of your genetic code. By learning how to influence your epigenome, you can become the architect of your own health. You'll no longer be just a reader of the book of life—you'll be the active author of your own masterpiece of well-being and longevity.

Through this book, I'll empower you to take control of your epigenetic health. I'll give you specific nutritional tools and strategies to help you decode your genes' secret language and rewrite your future using key foods and nutrients. You'll learn how to:

- Boost your SAMe levels: to slow aging and lower disease risk.
- Reduce homocysteine levels: a toxic byproduct of methylation.
- Enhance the activity of the enzymes that regulate acetylation: another core epigenetic process.

My goal is clear: to offer you accessible, practical knowledge on how to harness nutritional epigenetics to optimize your health and slow the aging process. As in my previous books, we'll begin with a solid theoretical foundation to understand the reasoning behind each intervention (*knowledge is never a burden*). Then we'll focus on what matters most—practice: how to apply what we've learned to lower disease risk, improve existing conditions, extend your healthy lifespan, and boost your physical and mental performance.

> **Theory without practice is useless, and practice without theory is dangerous.**

This is not just another book about nutrition; it's a guide that takes you deep inside your cells—where your present and future health are built. The time has come to nourish your genes for a longer, healthier, and more fulfilling life.

Let's begin.

THEORY

(Practice Without Theory Is Dangerous)

CHAPTER 1
What Is Epigenetics and How Does It Affect Life

Let's start with a technical definition: epigenetics is the study of changes that regulate the expression of DNA without altering its sequence. This field primarily focuses on two processes: (1) DNA methylation and (2) histone modification. In other words, epigenetics explores how habits and environmental factors can cause changes that affect how the information in DNA is interpreted—without modifying the DNA itself. Let's take a closer look.

In our biology, there are two types of information:

1. **Genetic**: This is the information stored in your DNA. It's what we inherit from our parents and it doesn't change—we're born with it and we die with it. It's a set of encoded "instructions" that guide our development and function; essentially, it's our *birth instruction manual.*

2. **Epigenetic**: This refers to the mechanisms that can change and regulate how genetic information is used. Epigenetics determines how DNA instructions are expressed in each cell—for example, which genes are silenced and which are activated. This process enables, among other things, *cell differentiation*— that is, how each cell (despite having identical genetic material) takes on one of thousands of different functions. For instance, an eye cell contains the same genetic information as a liver cell, but it's the epigenome that, by activating some genes and silencing others, makes the two cells different and suited to their roles. The epigenome adds and removes chemical "tags" on DNA, such as methyl groups (which silence genes and DNA regions) and acetyl groups (which activate them), without altering the DNA itself. Think of it like a vase with a post-it

note that says: "fragile, do not touch." The vase itself hasn't changed, but the added information helps preserve it and reduces the chances of it being damaged.

As Carlos López-Otín writes in his book *La vida en cuatro letras* (*Life in Four Letters*):

> The genome stores information, while the epigenome organizes it [...] the epigenome is something like the grammar or orthography of the genome [...] epigenetic changes act like accents, commas, umlauts, or periods that give grammatical meaning to the genetic message and reflect the dialogue between the genome and the environment in which life unfolds [...]. Thus, depending on our diet, body temperature, physical activity, or the emotions we experience, epigenetic changes occur—in the form of DNA methylation or histone modification—that determine whether the information in the genome is expressed or not, whether genes are turned on or off, whether they speak or remain silent.

In other words, epigenetics is how DNA information is interpreted depending on the changes it undergoes under the influence of environmental factors. As we'll see, this process governs much of our lives.

GENETICS VS. EPIGENETICS: WHICH IS MORE IMPORTANT?

Now that we understand what epigenetics is, a natural question arises: which is more important for our health—genetics or epigenetics? The answer isn't simple, and, as always, it depends on the context. Both play essential roles, but in different and complementary ways.

Genetics provides the starting point—the "instructions" we're born with—determining factors such as our predisposition to certain diseases. These aspects are fixed and don't change throughout life, though they influence many elements of our health and development.

Epigenetics, on the other hand, modulates how that genetic information is expressed, allowing it to adapt to our environment and the changes we experience. It's influenced by habits and environmental factors like nutrition, exercise, stress, sleep, age, emotions, and exposure to toxins. This means that, while we can't change our genetics, we can influence how it's expressed through the choices we make.

To better understand how habits, genetics, and epigenetics interact, let's use an analogy with two brand-new cars. One is a luxury vehicle—say, a Ferrari—and the other is more modest. Both roll off the factory line with their own unique features. The luxury car has a powerful engine and advanced technology. The simpler car also gets the job done, just without the flair.

In this analogy, genetics is the car's initial quality. The Ferrari represents "favorable" genetics—genes that offer advantages, such as lower disease risk or enhanced physical performance. The more modest car symbolizes less favorable genetics, with greater predisposition to illness or physical limitations.

This is where epigenetics comes in—it's the care and maintenance that affect how those cars perform. If the luxury car (our good genetics) is poorly maintained—never taken in for service, exposed to harsh conditions, or left damaged—it will have a shorter lifespan and poorer performance. On the other hand, if the modest car is meticulously cared for—regular checkups, stored in a garage, and kept in great shape—it can exceed expectations and even outlast the luxury model. The care and maintenance, which shape the car's performance, can end up being more important than its factory specs.

Likewise, even though good genetics may give us an advantage, it's our habits and environment that determine how those genes are expressed. A person with favorable genetics may still develop disease if they neglect their health—through poor diet, lack of exercise, insufficient sleep, or unmanaged stress. Meanwhile, someone with less favorable genetics but diligent health habits can live a long, healthy life. Just like a luxury car doesn't guarantee longevity without proper care, good genes alone aren't enough if we don't make the right choices to optimize how they function.

In short, both factors are essential to health:

- Genetics defines the range of possibilities—the limits we're born with. For example, if someone has genes that increase the risk of cardiovascular disease, those genes can't be changed.

- Epigenetics, however, gives us an opportunity to influence how those genes are expressed. Through healthy choices like proper nutrition and regular exercise, we can "press the right keys" to minimize the impact of harmful genes and enhance the benefits of protective ones.

There's no simple answer to which is more important, but epigenetics has a clear advantage: it's modifiable and responds to our decisions. While genetics sets the starting point, epigenetics allows us to either maximize or squander that potential. It gives us a measure of control over our health. It's the bridge between genes and the environment —and it allows us to compose our own *symphony* of health and well-being.

CORE EPIGENETIC PROCESSES

As we mentioned earlier, epigenetics regulates how genes are expressed through two main mechanisms: DNA methylation and histone modification. Let's take a closer look.

DNA Methylation

This process controls DNA activity—it silences the expression of a gene or a section of DNA to prevent it from being activated at the wrong time. It involves the addition of a chemical group called a methyl group to the DNA. The key compound for this process is the molecule S-adenosylmethionine (SAMe), which acts as the methyl group donor.

You could think of it as placing a mark or label on the DNA (the methyl group donated by SAMe) that tells the body not to activate the information in that region. If DNA is like an instruction manual, methylation is like placing a bookmark on certain pages so they won't be read.

There are two types of DNA methylation:

- **Global methylation**: Its main role is to stabilize the genome. Global methylation occurs in what are known as "non-promoter

and non-coding regions" of the DNA. These parts don't directly control gene activation or suppression. Instead, they act as *support areas* of the genome that don't contain key instructions for cellular function but still need to be kept under control so the DNA remains well-organized and protected. Methylation in these regions helps prevent these sequences from being mistakenly activated, which could cause genomic instability or mutations that may lead to diseases like cancer.
- **Specific methylation**: This form of methylation is essential for cellular differentiation. It occurs in "promoter regions" that regulate gene expression. By marking genes that should remain inactive, methylation ensures that only the genes needed for a particular cell's function are turned on. This is what allows cells to specialize and carry out specific roles in different tissues.

In the next chapter, we'll go deeper into the importance of SAMe and this process of methylation, because for everything to function properly, a gene must be expressed only when it's needed.

Just as, in my view, we humans should strive to do as well.

Histone Modification

Histones are proteins around which DNA is wrapped. They're crucial because human DNA is extremely long—if fully extended, each DNA molecule would measure about 2 meters (roughly 6.5 feet). Histones allow this long strand of DNA to be efficiently compacted so it can fit not just inside a microscopic cell, but inside its nucleus.

Isn't biology—and the wonders that evolution has achieved over millions of years of continuous refinement—truly spectacular? It has managed to fold nearly 2 meters of DNA "thread" into the nucleus of every one of our cells! The more I learn about biology, the more it continues to amaze me.

Back to histones: the key point is that certain chemical modifications to these proteins cause the DNA to become more or less tightly wound, which affects how accessible a gene is for being "read" or expressed. Only an unwound gene can be read. The more tightly packed the DNA, the less accessible it becomes.

These chemical changes, known as *post-translational modifications of histones* (PTMs), include several types: phosphorylation, ubiquitination, glycosylation, citrullination, and more. This variety is referred to as the *histone code*. Each modification can occur in different combinations and contexts, modulating gene expression in complex and specific ways. While all of them are important, this book focuses on two in particular: acetylation and deacetylation.

- **Acetylation** is a process in which enzymes called histone acetyltransferases (HATs) add a chemical group called an acetyl group to histones. This change relaxes the DNA, unwinding it and making it more accessible so that genes can be expressed. It's like loosening a knot, which makes it easier to unravel.
- **Deacetylation**, on the other hand, is the opposite process. Here, enzymes known as histone deacetylases (HDACs) remove those acetyl groups, causing the DNA to become tightly packed again. This silences genes or makes them harder to access—like tightening the knot again, restricting access to the DNA.

Unlike methylation, which acts directly on the DNA by placing a fixed label to prevent it from being read, deacetylation modifies the histones that wrap the DNA, compacting it and temporarily closing off access. Methylation is more rigid and long-lasting, while deacetylation is reversible depending on the cell's needs—more flexible and adaptive. Still, both processes occur in tandem and in harmony.

As we'll explore later, there are times when it's beneficial to *turn off* this gene-silencing mechanism. That's where natural HDAC inhibitors come into play—compounds that help *reactivate tumor suppressor genes*. Among these are epigallocatechin gallate (EGCG) from green tea, butyrate produced by gut bacteria when digesting resistant starch, selenium, sulforaphane from cruciferous vegetables, and other molecules we'll examine in detail in Chapter 11.

In short, we can think of DNA methylation as placing sticky notes on certain parts of the genome to tell the body not to "read" them, while histone modification is more like adjusting how accessible the DNA is for reading.

If we imagine the body as a large company, methylation would be like the long-term policies set by upper management: once established, they're durable and guide the company's behavior over time. Acetylation and deacetylation, meanwhile, are like the employees—operating within those policies but quickly adapting to the demands of each day, responding to changing circumstances. While management's policies are stable, the workers are flexible; both are essential to the company's success—or, in our analogy, to the cell's ability to respond and adapt to environmental changes.

As we know, evolution doesn't favor the strongest species, but the ones best able to adapt to the ever-changing world around them.

Histone Lactylation: The Link to Metabolism

In addition to the better-known mechanisms like acetylation and deacetylation, a new type of histone modification was discovered in 2019: *lactylation*.[1] Since this process also modifies histones, it could have been included in the previous section along with acetylation and deacetylation. However, I've chosen to dedicate a separate section to it, because it's a recent discovery and still not as well understood or extensively studied as the others.

What makes this modification particularly interesting is its direct connection between how cells generate energy (cellular metabolism) and how genes are regulated—something that hadn't been clearly observed with other epigenetic modifications until now. Let's take a closer look.

Lactylation involves the addition of a chemical group derived from lactate—known as a *lactyl group*—to histones. This process increases the accessibility of DNA for gene expression, much like acetylation does. This implies that lactate, a molecule once thought to be merely a waste product of metabolism (produced when the cell generates energy from glucose), plays a far more important role than previously believed. We now know that lactate, far from being just a metabolic byproduct, is essential to many biological processes.

---

[1] https://doi.org/10.1038/s41586-019-1678-1

Lactate: From Waste to Key Player

Lactate is no longer viewed merely as a metabolic byproduct. In fact, it's now recognized as a key player in cellular function. For example, it serves as a metabolic intermediate and is directly used as a source of energy by organs such as the heart, brain, and muscles. Additionally, through the Cori cycle, lactate can be converted back into glucose, making it a recyclable energy source.

But its role goes beyond energy. Recent studies have shown that lactate also acts as a signaling molecule, activating genes through histone lactylation. This impacts a wide range of biological processes, including inflammation, wound healing, memory, and mitochondrial production.[2]

How Does Lactylation Work?

The mechanism of lactylation is very similar to the processes of acetylation and deacetylation. Enzymes such as p300 (a type of HAT) act as "writers," adding lactyl groups to histones, which "opens" the DNA so that certain genes can be read and expressed. In contrast, enzymes like HDACs, which are involved in deacetylation, act as "erasers," removing those lactyl groups.[3] This means that lactylation is a reversible process and can be adjusted based on the cell's needs.

This discovery offers a new perspective on how cells not only respond to their external environment, but also sense their own internal metabolic state—adding yet another layer of complexity to epigenetic regulation.

Lactate, Exercise, and Cancer

Lactylation is especially relevant because it shows that a cell's metabolic state can directly influence epigenetic regulation. Depending on how much lactate is produced, certain genes can be activated or silenced—something that's crucial in diseases like cancer, where cellular metabolism is profoundly altered. But this mechanism also has implications in other contexts, such as inflammation and physical exercise.

---

[2] https://doi.org/10.1113/JP280955
[3] https://doi.org/10.1186/s13148-024-01682-2

Physical exercise has been shown to play a protective role in cancer prevention and to improve outcomes in diagnosed patients through various mechanisms;[4,5] one of which appears to involve lactate.

Lactate's role in cancer is complex. Cancer cells tend to rely primarily on anaerobic glycolysis—a process that allows them to derive energy from glucose without depending on oxygen, generating lactate as a byproduct. What's particularly interesting is that they prefer this method even when oxygen is available—a phenomenon known as the *Warburg effect*.

Normally, in the presence of oxygen, cells generate energy through the mitochondria via oxidative phosphorylation, which is more efficient because it produces more energy (adenosine triphosphate [ATP]), though not as quickly as glycolysis. It was traditionally believed that cancer cells relied on glycolysis because their mitochondria didn't function properly. However, recent studies have shown that this isn't the case: the mitochondria in cancer cells are fully functional. What happens is that these cells *choose* glycolysis because they need lactate to build the "building blocks" (DNA, lipids, and proteins) required for rapid growth and division.

So, cancer cells use lactate not only as an alternative energy source, but also as a signal to support their survival and proliferation.

Knowing this, we can say that the problem isn't the production of lactate itself, but rather the body's inability to efficiently clear it.[6] When lactate builds up uncontrollably, it creates an altered metabolic environment—acidic and pro-inflammatory—that promotes tumor growth, damages surrounding tissue, and suppresses the immune response.

This is where physical exercise comes in—it helps protect the body against cancer through two pathways: one short-term, during the exercise itself, and another medium- to long-term, as the body adapts to the stimulus of physical exertion.

---

[4] https://doi.org/10.1093/bmb/ldab019
[5] https://doi.org/10.1002/cac2.12488
[6] https://doi.org/10.1093/carcin/bgw127

In the short term, the lactate produced during exercise is used by active muscles as a source of energy, reducing its availability to cancer cells. These cells, which depend on lactate for their growth and survival, may be constrained by having to compete with healthy cells for this resource during exercise.

In the medium and long term, the lactate generated during exercise can trigger histone lactylation, promoting the creation of new mitochondria by activating genes such as *PGC-1α*.[7] Since mitochondria also consume lactate, this increase in mitochondrial mass enhances the body's ability to eliminate lactate, reducing resting lactate levels and its availability to cancer cells—thereby limiting their proliferation.

Thus, by improving mitochondrial biogenesis and optimizing energy metabolism, regular exercise not only enhances overall health but may also serve as a protective strategy against cancer by reducing the chronic buildup of lactate in the body.

EPIGENETIC ALTERATIONS: IMPACT ON HEALTH AND AGING

Epigenetic alterations are one of the four primary causes of aging among the nine hallmark characteristics identified in 2013 by Carlos López-Otín and other researchers in one of the most influential scientific publications in history: *The Hallmarks of Aging*.[8] In 2023, that list was updated to include twelve hallmarks.[9] One key driver behind these alterations is a malfunction in DNA methylation and the gradual loss of its capacity—leading to the erosion of epigenetic information.

Let's explore this in more detail, because understanding the broader context is essential if we want to learn how to slow down aging by enhancing our methylation capacity as it naturally declines over the years.

Another analogy can help us grasp the importance of epigenetics in health and aging. Imagine our genomic system as a piano and a pianist. The piano represents our genome—the genes we're born

---

[7] https://doi.org/10.1096/fj.07-8174com
[8] https://doi.org/10.1016/j.cell.2013.05.039
[9] https://doi.org/10.1016/j.cell.2022.11.001

with—while the pianist is our epigenome, the one who decides how those genes are expressed, whether they're activated or kept silent (in other words, whether or not the keys are pressed to make them sound). We may have a piano of higher or lower quality, but it's the pianist who plays the keys and interprets the melody—determining how that piano sounds, for better or worse.

> **A poor-quality piano in the hands of an excellent pianist will always sound better than a high-end piano played by a terrible musician.**

Epigenetic information can be modified relatively easily in response to the internal or external demands of the cell's environment—the pianist adapts to the setting, interpreting the melody one way or another as circumstances require. But this adaptability comes at a cost: unlike genetic information (which we inherit and which remains unchanged throughout life—the piano is always the same), epigenetic information deteriorates and changes as we age—the pianist gradually loses skill with time.

Let's break this down further.

When a cell ages or becomes damaged, it needs to be replaced. To do that, it copies its "instruction manual" (the DNA) and divides in two, creating a new cell with its own "manual." It's during this copying process that epigenetic information is lost—one of the main causes of the body's deterioration as we age. To put this into perspective: tens of trillions (with a *t*) of cells are renewed every day, so you can imagine the scale of this process day after day, year after year.

To make matters worse, these epigenetic alterations also accelerate the shortening of telomeres—another of the original nine hallmarks of aging. Telomere shortening is often compared to the fraying of aglets, the plastic tips at the ends of shoelaces that protect them from unraveling. Telomeres are the regions at the ends of chromosomes (the structures within the cell nucleus that hold DNA), and they serve to protect genetic information during cell division. With each division, these tips wear down a bit, becoming shorter. When they get too short, they can no longer properly protect the chromosomes, which can lead to genetic instability, errors in cell division, and contribute to aging and age-related diseases.

KEY POINTS

1. **Definition of Epigenetics**: It studies how environmental factors and lifestyle habits influence the way the information in our DNA is interpreted—without altering the DNA itself.

2. **Difference Between Genetics and Epigenetics**:
   - Genetics defines our starting point. It's the "instruction manual" for our body that we inherit.
   - Epigenetics regulates how and when that genetic information is expressed—what parts of the manual are read or ignored—and is influenced by factors like environment, diet, exercise, sleep, and stress.
   Genetics is fixed, but epigenetics is modifiable.

3. **Car and Habits Analogy**: Genetics is the initial quality of the car, while epigenetics is how it performs based on the care it receives. Good maintenance can allow a modest car to outperform a poorly maintained luxury car.

4. **Environmental Impact on Genes**: Life choices like diet and exercise influence gene expression, helping reduce the impact of harmful genes and enhance the effect of beneficial ones.

5. **Core Epigenetic Processes**:
   - **DNA Methylation**: Silences parts of the DNA, such as genes that don't need to be active, by adding chemical "tags" called methyl groups. SAMe is key in this process as the donor of those methyl groups. There are two types of methylation based on the region of the DNA involved:

- **Global**: Occurs in non-promoter regions that don't control gene activity. Helps stabilize the genome.
- **Specific**: Occurs in promoter regions that regulate gene expression. Supports cell differentiation and specialization.
  - **Histone Modification**: Regulates DNA accessibility. Several processes exist, but in our context, three are particularly important:
    - **Acetylation**: Opens the DNA, making it easier to read and express. The enzymes involved are HATs.
    - **Deacetylation**: Recompacts the DNA, restricting access and gene expression. The enzymes involved are HDACs.
    - **Lactylation**: A recently discovered process that also modifies histones and links cellular energy to epigenetics. Lactate, a byproduct of glucose metabolism—especially during exercise—can activate genes through lactylation, affecting processes like inflammation, memory, and mitochondrial production.

6. **Epigenetics, Exercise, and Cancer**: Cancer cells rely on lactate to grow, but physical exercise reduces lactate availability by activating genes that promote mitochondrial production. Mitochondria consume lactate and help clear it, potentially limiting tumor growth.

7. **Epigenetics and Aging**: Epigenetic alterations play a key role in aging. As we get older, the body's ability to maintain proper epigenetic marks (like methylation) declines, contributing to genomic instability and accelerating the loss of cellular information—leading to damage and aging.

8. **Piano and Pianist Analogy**: Genetics is the "piano" and epigenetics is the "pianist." Regardless of how good the piano is, it's the pianist who determines how the music is played. A skilled pianist can make any piano sound beautiful—so we should try to preserve the pianist's ability over time. Genetics sets the potential; epigenetics allows us to either realize or waste that potential.

**

Now then, after everything we've covered—and recognizing the challenges involved in modifying epigenetic changes—some important questions arise: Is it possible to intervene in these changes through nutrition? Can we slow down the loss of our ability to maintain epigenetic marks, such as methylation, that contributes to aging?

The answer is encouraging. In 2023, it was shown that many of these alterations are largely reversible.[10] How we can nutritionally influence these changes will be explored in the practical section of this book. But before we get there, it's important to meet one of the main protagonists of this story: S-adenosylmethionine, better known as SAMe.

---

[10] https://doi.org/10.1016/j.cell.2022.12.027

CHAPTER 2
The Role of SAMe in Epigenetics

In the previous chapter, I explained that one of the two main epigenetic processes is DNA methylation—a mechanism through which methyl groups (a kind of "tag") are added to the DNA. These tags are donated by the molecule S-adenosylmethionine (SAMe) and signal to the body which areas of the DNA and which genes should not be expressed.

As the years go by and we age, our methylation capacity declines. This makes it harder to maintain those epigenetic marks, leading to a loss of cellular information, destabilization of the genome, and, consequently, poorer cell function. This contributes to aging and increases the risk of many diseases, which we'll explore in Chapter 5.

One of the goals of this book is to explore how we can slow down this decline by enhancing SAMe availability.

Let's take a closer look.

WHAT IS METHYLATION AND THE ROLE OF SAMe AS A METHYL GROUP DONOR

Methylation is a biological process essential to life that takes place inside cells. In this process, the molecule SAMe transfers a methyl group ($-CH_3$: one carbon atom and three hydrogen atoms) to other biomolecules such as DNA, proteins, lipids, and hormones, among others.

SAMe is the second most-used molecule by enzymes, right after adenosine triphosphate (ATP), which we need to produce energy. Enzymes are vital for life because they enable and regulate nearly all biochemical reactions in living organisms. In other words, without

SAMe, many chemical reactions essential to life could not occur—or would be severely impaired.

Understanding this process is key to grasping the concepts we'll explore next in this book, so let's take a moment to be sure we understand it well.

Think of SAMe as a delivery service moving through your body, carrying special "packages" (methyl groups) that act like small "tags" delivered to different molecules. When a molecule receives one of these tags, it can change its behavior and function differently—or even transform chemically into another molecule. Although this tag doesn't alter the molecule's basic structure or identity, it does modify how it functions and how it interacts with other molecules.

For example:

- When DNA receives a methyl group from SAMe, it's still DNA—but its behavior, or how a specific gene is expressed, may change.
- A methylated protein is still the same protein, but its activity or its interaction with other molecules may be different.

It's like adding a small accessory to an object: the object remains the same, but its function may be slightly altered—or it may now carry important new information.

When SAMe donates its methyl group, methylation occurs—and not only does the recipient molecule change, but SAMe itself is also transformed into S-adenosylhomocysteine (SAH), which then becomes homocysteine. Due to its importance to health, we'll explore this process in more detail in Chapter 4.

SAMe AND DNA METHYLATION

Among all the functions that require SAMe for methylation (which we'll explore in the next chapter), the one most relevant to epigenetics is DNA methylation—one of the two main epigenetic mechanisms, as we saw in the previous chapter.

We've already said that DNA is like a vast "instruction manual" containing all the information needed to build and operate the body. That manual is written using just four letters, known as "nitrogenous bases": adenine (A), guanine (G), cytosine (C), and thymine (T). Cytosine receives the methyl group from SAMe, forming 5-methylcytosine (5mC)—a kind of "tag" that marks which parts of the manual should not be read. This marker is added with the help of the enzyme DNA methyltransferase (DNMT).

Let's return to the idea of the human body as a large company. In this analogy, DNA is the "operations manual." DNA methylation acts as a "labeling system" for this manual, placing sticky notes (the methyl groups) that tell each cell which parts of the manual to ignore or which genes not to activate. These notes allow the company to operate without errors, ensuring that only the necessary instructions are used at the right time.

SAMe, in this case, is the "supplier" that provides the raw material needed to produce those sticky notes. If there's enough SAMe, the labels are applied correctly, and the company (the body) can function efficiently. But if there's a shortage of SAMe, some labels will be missing, and cells may end up reading parts of the manual they shouldn't —leading to serious mistakes in the system, like malfunctioning processes or major disruptions comparable to disease.

It's also important to remember that each cell in the body contains a full copy of this manual—that is, every cell holds the same complete set of instructions, with more than 20,000 genes and other relevant information. Thanks to methylation, a cell in the eye, for example, reads only the instructions relevant to being an eye cell and ignores those for becoming a kidney cell. This process is essential for development and cellular differentiation, allowing each cell type to specialize and perform its specific function properly—ensuring that the body operates in an orderly and efficient way while preserving the integrity and stability of the genome.

Now imagine how that company would run if there were no specialization and everyone worked in IT.

SAMe AND AGING

At the end of the previous chapter, I mentioned some good news: epigenetic alterations are reversible. SAMe availability decreases with age, and this decline isn't linear—it's exponential, accelerating as we grow older.[11] However, it is possible to slow down this drop and increase SAMe levels through what we might call "nutritional intervention." How can we do that? In two main ways:

1. Ensuring that homocysteine doesn't get stuck in the system and can be recycled to generate more SAMe.
2. Conserving SAMe—meaning, providing the body with certain molecules exogenously, through food and supplements, that normally consume large amounts of SAMe when synthesized endogenously. This way, we avoid wasting SAMe and keep it available for other important functions related to epigenetics and aging.

We'll explore in more detail how to apply these principles in the practical section of the book, so don't worry. For now, let's look at how a shortage of SAMe causes epigenetic alterations—one of the keys to aging, as we've seen, and to many diseases we'll analyze in Chapter 5.

Epigenetic Drift

A decline in SAMe—whether due to aging, genetic variants, or poor nutrition—leads to altered DNA methylation and an overall deterioration of epigenetic markers. In other words, as we age, we lose pieces of information the body needs to function properly.[12] This process, known as *epigenetic drift*, contributes to:

1. The progressive decline of cellular functions, accelerating the aging process.

---

[11] https://doi.org/10.1073/pnas.1120658109
[12] https://doi.org/10.2217/epi-2016-0078

2. Impaired cellular repair, which makes cells less adaptable as we age—they can no longer respond or adjust to changes the way they did in youth.

Technically, this is referred to as a *loss of phenotypic plasticity*—the diminished ability of a cell or organism to adapt and alter its characteristics (both physical and functional) in response to changing environments, conditions, or stimuli.

The alteration of the epigenomic state involves two main processes:

1. **Global hypomethylation of the genome**: As we age, the decline in SAMe leads to reduced DNA methylation across the genome—especially in non-promoter regions (outside CpG islands), meaning parts of the DNA that don't directly control genes. As we saw in the previous chapter, the absence of "tags" in these areas can disorganize the DNA and promote genomic instability.

    This is actually a protective mechanism. DNA contains repeated sequences and mobile genetic elements that, if activated, could cause mutations and disrupt cellular function —increasing the risk of cancer and other diseases as cells grow older. These repetitive elements and mobile sequences are like "pages in the instruction manual that should not be read," because doing so could cause confusion and errors in the instructions. Methylation places the "tags" (supplied by SAMe) on those sections to keep them from being read (or activated). Without enough SAMe, there are no "tags."

2. **Specific hypermethylation of certain genes**: In some key genes—such as tumor suppressor genes and others involved in cellular repair (like telomerase)—excessive methylation builds up in the promoter regions (CpG islands) that regulate gene activity.

    This over-tagging silences those genes, preventing them from doing their job in situations where they should be active, which increases the risk of disease development. This phenomenon is known as *epigenetic silencing*. In this case, hypermethylation is part of the overall aging process and isn't necessarily caused directly by SAMe deficiency.

Returning to the analogy of DNA as a piano: each key represents a part of the DNA (such as genes or repetitive elements), and the pianist (epigenetics) decides which keys to press—or not press—at any given moment in order to play the best possible melody. 5mC, which comes from the methyl groups donated by SAMe, acts like a mark on a piano key telling the pianist not to play it, so it doesn't sound when it shouldn't. You could say that 5mC is a control system that ensures the right parts of the DNA are used at the right time, so the music (the body's functioning) produced by the piano (DNA) remains harmonious.

So, if we don't have enough SAMe (global hypomethylation), the body can't produce 5mC, which promotes *epigenetic drift*—a slow, gradual degradation in how the DNA's instructions are read as we age. The DNA itself doesn't change, but its interpretation becomes increasingly flawed. This gives us a clearer sense of how important it is to prevent the decline of SAMe with age, in order to preserve our methylation capacity.

On the other hand, imagine that in other parts of the piano, some keys become stiff—or even locked—because of excessive tagging (specific hypermethylation of certain genes). This prevents the pianist from playing those keys properly. As a result, the music can't be played when it's needed—in other words, it prevents certain genes from being expressed at the right time. A good example is tumor suppressor genes, whose vital role can be compromised by this process.

In summary, you can think of epigenetic drift as the progressive deterioration of how the piano (the genome) is played, caused by changes in how the pianist (epigenetic mechanisms) interacts with the keys (different parts of the DNA). Over time, the pianist has a harder time playing certain keys effectively, and this can distort the music (the body's functioning), making it lose its harmony and beauty.

> **KEY POINTS**

1. **Methylation and the Role of SAMe:**
   - **Methylation**: A process in which a small chemical "tag" called a methyl group is added to biomolecules, including DNA. This tag can alter how a molecule functions without changing its structure.
   - **Function of SAMe**: SAMe is essential to this process, as it donates these methyl groups—acting as both a "supplier and delivery service" of the raw material needed for methylation. Methylation occurs when SAMe transfers a methyl group.

2. **SAMe and DNA Methylation:**
   - **DNA Methylation**: Among the many roles of methylation, the one relevant to epigenetics is DNA methylation—our "instruction manual." SAMe donates methyl groups to DNA, forming 5mC, a "tag" that tells the cell which genes and DNA regions to ignore. This process is key to:
     - **Genome stability**: When methylation is global (in non-gene-controlling regions), it prevents the activation of repetitive or mobile DNA elements that could disrupt proper gene interpretation and cause errors.
     - **Cellular differentiation**: When methylation is specific (in gene-regulating regions), it ensures that cells only read the instructions necessary for their specific role, enabling each cell type to specialize.

3. **SAMe and Aging:**
   - **SAMe Decline**: As we age, SAMe levels drop, impairing the body's ability to maintain epigenetic tagging and disrupting genomic balance—contributing to aging and disease risk.
   - **Nutritional Interventions**: This loss can be slowed by increasing SAMe levels—either by managing homocysteine recycling or supplementing with molecules that otherwise consume SAMe. These strategies will be covered in the practical section.

- **Epigenetic Drift**: This is the age-related loss of epigenetic tags. It not only affects cell function and repair, but also accelerates aging. It includes two processes:
  - **Global hypomethylation**: With age, a lack of SAMe reduces methylation in non-coding DNA regions. These areas lose the "tags" that maintain order and genomic stability—leading to disorganization and higher disease risk.
  - **Specific hypermethylation**: Occurs in key genes like tumor suppressors, where excessive tagging blocks their protective functions—raising the risk of diseases such as cancer.

4. **SAMe in the Piano Analogy**:
   - SAMe provides the "tags" that mark the piano keys (DNA regions), guiding the pianist (epigenetics) on which keys to play and which to avoid. Without these marks, the pianist makes mistakes in the music (the body's functioning).

5. **Conclusion**:
   - DNA methylation is crucial for healthy aging and disease prevention. Maintaining adequate levels of SAMe is essential for proper gene regulation and to avoid age-related dysfunctions.

CHAPTER 3
Other Functions of SAMe and Methylation

In the previous chapter, we discussed how S-adenosylmethionine (SAMe) plays a key role in DNA methylation—an essential process for regulating gene expression and maintaining the stability of our genome. However, the function of SAMe and methylation goes well beyond this epigenetic regulation. This small molecule is also a fundamental component in many other biochemical reactions that sustain cellular balance and overall health.

Always acting as a methyl group donor, SAMe is involved in other methylation processes ranging from the synthesis of creatine—vital for muscle function—to the production of neurotransmitters such as serotonin and dopamine, which are essential for emotional well-being. It also plays a role in regulating circadian rhythms, breaking down catecholamines and estrogens, and producing antioxidants that protect cells from oxidative damage.

In this chapter, we'll take a deeper look at these additional functions of methylation. We'll see how SAMe's capacity to donate methyl groups affects key physiological and metabolic processes that support our well-being—far beyond genetic control.

CREATINE SYNTHESIS

This process begins in the kidneys, where the enzyme L-arginine:glycine amidinotransferase (AGAT) uses glycine and arginine to form guanidinoacetate, the precursor to creatine. Then, in the liver, SAMe donates its methyl group to convert guanidinoacetate into creatine, with the help of the enzyme guanidinoacetate N-methyltransferase (GAMT).

Creatine is essential for energy production and management in cells—especially in tissues with high energy demands, such as muscles and the brain.

- **In muscle**: Creatine acts as a rapidly available energy reserve. It's stored as phosphocreatine, which can quickly donate its phosphate group to convert adenosine diphosphate (ADP) into adenosine triphosphate (ATP)—the cell's main energy source. This process enhances strength, endurance, and muscle recovery. We'll explore this topic further in an appendix at the end of the book.
- **In the brain**: Creatine also plays a vital role in brain energy metabolism, supporting ATP production and improving mitochondrial efficiency—mitochondria being the "powerhouses" of the cell. This is particularly important in neurons, which have high and fluctuating energy demands. Additionally, creatine has antioxidant and neuroprotective properties; it also supports the development of new brain cells by promoting the growth and maturation of neural precursor cells—essential for brain tissue regeneration and repair.

PHOSPHATIDYLCHOLINE SYNTHESIS

Phosphatidylcholine is an essential component of the body, and its synthesis depends on methylation. This process begins when phosphatidylethanolamine receives three methyl groups from SAMe, converting it into phosphatidylcholine. The enzyme that catalyzes this transformation is phosphatidylethanolamine N-methyltransferase (PEMT).

Phosphatidylcholine is indispensable for several vital functions in the body:

- **Structural component of cell membranes**: Phosphatidylcholine is a key element in cell membranes, where it contributes to membrane fluidity and integrity—fundamental for proper cellular function. In the brain, its role is especially critical, as it

enhances neuronal function and offers protection against neurodegenerative disorders such as Alzheimer's disease.
- **Fat metabolism in the liver**: Phosphatidylcholine plays a crucial role in fat metabolism, helping prevent fat accumulation in the liver—a condition known as hepatic steatosis. It also facilitates the transport and metabolism of cholesterol and other lipids.
- **Formation and secretion of very-low-density lipoproteins (VLDL)**: In the liver, phosphatidylcholine is essential for forming and secreting VLDL, which are crucial for lipid transport throughout the body.
- **Precursor to choline**: Phosphatidylcholine also serves as a precursor to choline, a molecule with multiple functions. Choline, in turn, is a precursor to two other important compounds:
  - **Acetylcholine**: A key neurotransmitter involved in memory, learning, focus, and muscle control.
  - **Trimethylglycine (TMG)**: As we'll see later, TMG helps reduce homocysteine levels, supporting further SAMe synthesis. Additionally, as an osmolyte, TMG helps cells maintain volume, hydration, and stability.

REGULATING CIRCADIAN RHYTHMS

The enzyme Mettl3 transfers methyl groups from SAMe to adenosine on messenger RNA (mRNA)—a process known as mRNA methylation—which plays a fundamental role in regulating circadian rhythms, among other things.

The term *circadian* comes from the Latin *circa* (around) and *dies* (day), so it literally means "around a day." It describes the biological rhythms and physiological processes that follow a roughly 24-hour cycle, mirroring the day-night cycle of our planet. These circadian rhythms are crucial for regulating a wide range of biological functions, such as sleep, hunger, hormone secretion, body temperature, and other cellular activities.

We can think of circadian rhythms as the conductor of an orchestra, keeping the body in sync with the rhythm of day and night. However, in industrialized societies, many people experience disruptions to these rhythms because our lifestyle often conflicts with our biological needs: we avoid early-morning natural light at low angles, we don't get enough bright light during the day, and instead we're exposed to artificial light at night. This misalignment affects not only sleep, but also mood, as we'll explore later in the section on depression in Chapter 5.

Roughly 10% of our genes are regulated by a "master clock" located in the suprachiasmatic nucleus, a structure within the hypothalamus just above the point where the optic nerves cross (the optic chiasm). This "clock," which governs so many genes, depends on the light that enters through our eyes.

If DNA is like an instruction manual containing all the genetic information needed to build and maintain an organism, then mRNA —which receives a methyl group from SAMe—is like a specific copy of one section of that book. This mRNA carries genetic instructions from DNA to the ribosomes, the cell's "protein factories." In other words, mRNA is used to produce proteins according to the cell's needs.

When mRNA methylation is impaired due to low availability of SAMe, the master circadian clock is significantly disrupted. As a result, the regulation of metabolism and basic biological functions is affected.[13] This can cause the circadian cycle to lengthen beyond the usual 24 hours—and the less methylation there is, the longer this cycle becomes. The consequences of this misalignment include a cascade of problems: sleep disorders, hormonal imbalances, fatigue, mood changes, cognitive issues, increased risk of overweight, diabetes, cardiovascular disease, and more.

This link between methylation and the circadian clock is so fundamental that it has been preserved across more than 2.5 billion years of evolution.[14]

BREAKING DOWN EXCESS CATECHOLAMINES AND ESTROGENS

Catecholamines are a group of chemical substances that act as hormones or neurotransmitters; the three main ones are dopamine,

---

[13] https://doi.org/10.1016/j.cell.2013.10.026
[14] https://doi.org/10.1038/s42003-020-1031-0

norepinephrine (also known as noradrenaline), and epinephrine (also known as adrenaline). These substances play a crucial role in the stress response, mood regulation, blood pressure, and heart rate.

In situations of prolonged stress—due to certain medical conditions or genetic predispositions—the body can produce or release an excess of catecholamines. This may lead to symptoms such as anxiety, high blood pressure, rapid heartbeat, mental health issues, and other disorders. Therefore, proper regulation of catecholamines is essential for the health of both the nervous and cardiovascular systems.

Specific enzymes use SAMe, which donates its methyl group, to transform excess catecholamines into forms that the body can more easily eliminate—namely, into metabolites. One of the most well-known enzymes involved in this process is catechol-O-methyltransferase (COMT). COMT uses the methyl group from SAMe to methylate dopamine, converting it into 3-methoxytyramine, a metabolite that is an intermediate step in the breakdown and elimination of dopamine.

This same enzyme, COMT, also uses the methyl group from SAMe to convert catechol estrogens (a form of estrogen) into more water-soluble compounds. This transformation makes it easier for the body to excrete these estrogens and helps maintain proper hormonal balance.

BREAKING DOWN EXCESS HISTAMINE

Histamine is a chemical substance that plays a vital role in the immune system, digestion, and the central nervous system. It's best known for its role in allergic reactions—such as sneezing, itching, and skin redness. However, in some cases, the body can accumulate too much histamine, either due to excessive production or insufficient breakdown. This buildup can trigger allergy-like symptoms and contribute to conditions such as migraines, digestive issues, and sleep disturbances.

The enzyme histamine N-methyltransferase (HNMT) uses methyl groups provided by SAMe to convert histamine into N-methylhistamine. This process is essential for keeping histamine levels in check and preventing exaggerated allergic responses. It also helps reduce the risk of migraines and supports the proper function of the immune system and the body's inflammatory response.

Regulating histamine levels is so important that the body also relies on an additional enzyme to break it down: diamine oxidase (DAO). Unlike HNMT, which depends on SAMe and works systemically and at the cellular level, DAO degrades histamine in the gastrointestinal tract. This enzyme helps break down histamine ingested with food or released in the gut, preventing it from being excessively absorbed into the bloodstream.

BREAKING DOWN EXCESS NIACINAMIDE (VITAMIN B3)

Niacinamide, also known as nicotinamide, is a form of vitamin B3. It can be obtained through food—especially from animal sources—but it's also produced within the body as a byproduct of using nicotinamide adenine dinucleotide (NAD+).

NAD+ is a crucial coenzyme that binds to certain enzymes, helping them carry out vital chemical reactions in the body—especially in energy metabolism and DNA repair. You could think of NAD+ as the "fuel" that powers these reactions, and as it's used, niacinamide is released as a byproduct.

An excess of niacinamide is associated with problems such as insulin resistance, type 2 diabetes, neurological toxicity, obesity, and increased oxidative stress, among others. In other words, an accumulation of niacinamide in the body can be harmful. To prevent these issues, the body uses the enzyme nicotinamide N-methyltransferase (NNMT), which takes a methyl group from SAMe and attaches it to niacinamide—transforming it into a molecule that can be excreted in the urine.

It's essential to eliminate excess niacinamide when it accumulates. Additionally, NNMT has other beneficial effects, such as regulating lipid levels in the liver (by suppressing triglyceride and cholesterol synthesis when needed) and promoting antithrombotic and vasodilatory effects in the endothelium—the inner lining of blood vessels. However, ideally, niacinamide buildup should be avoided—not only to prevent the issues mentioned above but also to avoid unnecessarily depleting SAMe, which would reduce its availability for other critical functions.

Moreover, excess niacinamide and the resulting overexpression of NNMT have been observed in a variety of cancers, where they are associated with tumor progression, metastasis, and poorer survival rates. This is likely due to high niacinamide levels and the depletion of SAMe, which is diverted toward eliminating excess niacinamide.[15,16]

NEUROTRANSMITTER SYNTHESIS (SEROTONIN, DOPAMINE, AND NOREPINEPHRINE)

Neurotransmitters such as serotonin, dopamine, and norepinephrine play fundamental roles in both brain and bodily functions.

- **Serotonin**: Essential for regulating mood, emotional well-being, sleep, body temperature, and appetite.
- **Dopamine**: Closely associated with the brain's reward and pleasure system, it plays a crucial role in motivation and enthusiasm.
- **Norepinephrine (or noradrenaline)**: Involved in the body's "fight or flight" response, it regulates blood pressure and heart rate under stress and also influences attention, focus, and emotional responses.

Serotonin is synthesized from the amino acid tryptophan, while dopamine and norepinephrine are synthesized from the amino acid tyrosine. These synthesis processes involve a series of chemical reactions, in which the rate-limiting step—the slowest step that determines the speed of the reaction—depends on two enzymes that require SAMe as a methyl group donor cofactor.

In addition, SAMe has been shown to enhance the binding of these neurotransmitters to their receptors by increasing the fluidity of brain cell membranes. This effect is particularly important in conditions like Alzheimer's disease and depression, where reduced membrane fluidity has been observed.

---

[15] https://doi.org/10.1016/j.tem.2017.02.004
[16] https://doi.org/10.1016/j.molmet.2021.101165

SYNTHESIS OF THE HORMONE AND NEUROTRANSMITTER ADRENALINE

Adrenaline, also known as epinephrine, is a versatile molecule that can act either as a hormone in the endocrine system—coordinating the body's response to stress—or as a neurotransmitter in the nervous system, depending on where it is released.

Adrenaline is synthesized by the enzyme phenylethanolamine N-methyltransferase (PNMT), which uses SAMe as a source of methyl groups to convert norepinephrine into adrenaline. This process is essential for adrenaline to carry out its many functions in the body:

- Regulating blood pressure and respiration.
- Stimulating the release of hormones such as luteinizing hormone and growth hormone from the pituitary gland.
- Regulating circadian rhythms.
- Supporting cognitive processes such as learning, memory, attention, and emotional response.
- Participating in neurodegenerative processes, including those seen in Alzheimer's and Parkinson's diseases.[17,18]

SUPPORTING THE DIFFERENTIATION AND PROLIFERATION OF T LYMPHOCYTES IN THE IMMUNE SYSTEM

The adaptive immune system is a fundamental part of the body's "military defense system" and consists of two main types of "soldiers": T lymphocytes (helper and killer T cells) and B lymphocytes. These lymphocytes are a type of white blood cell (leukocyte) produced in the bone marrow. Let's look at how they function—and how SAMe fits into the picture.

When a virus, bacterium, or any other foreign substance enters the body, it presents antigens on its surface—unique "ID tags." The immune system recognizes these antigens as signals that something

---

[17] https://doi.org/10.1016/S0969-2126(01)00662-1
[18] https://doi.org/10.1021/jacs.0c05446

doesn't belong. Macrophages are the cells that patrol the body in search of these "invaders." They act like "garbage collectors": once they find an invader, they "engulf" it, break it down, and present its antigens on their surface to alert other immune cells, such as T cells, that there's a threat to eliminate.

To recap: macrophages identify, collect, and tag "invaders" with antigens so the immune system can recognize them. That's when T cells—or helper T cells, to be specific—step in. These are specialized "soldiers" that detect antigens using their T cell receptors (TCRs), a type of "sensor" that recognizes pathogen antigens. When a helper T cell's TCR binds to its specific target, the T cell is activated and sends signals to initiate the next immune response—activating both B cells and killer T cells.

At this stage, SAMe plays a crucial role. During the activation of helper T cells, SAMe provides the methyl groups needed for the epigenetic reprogramming of these cells—a process that modifies their "behavior" by altering their cellular programming.[19] This reprogramming enables T cells to proliferate (increase in number) and differentiate (specialize into specific subtypes). Through these two actions, the immune system builds a large, specialized response team that can effectively fight the infection or threat. Without SAMe, this process would not be possible.

Helper T cells don't just recognize invading pathogens—they also detect damaged, infected, or cancerous cells within the body. Cancer cells often display certain proteins or markers on their surface that are not typical of normal cells, and helper T cells, using their TCRs, can recognize these differences—flagging them so killer T cells can attack. Cancer cells go to great lengths to avoid detection, and cancer immunotherapy specifically aims to enhance the ability of T cells to recognize and destroy these cells more effectively.

Let's continue the process. Once activated, helper T cells send signals to activate both B cells and killer T cells. B cells then begin producing specific antibodies—another kind of "tag"—against the pathogen. These antibodies bind to the pathogen, marking it for

---

[19] https://doi.org/10.7554/eLife.44210

destruction. At this point, killer T cells step in to attack and destroy both infected cells and cancerous cells.

After an infection, some T and B cells become "memory cells." These cells "remember" the specific pathogen that caused the infection. If the same pathogen tries to invade the body again in the future, these memory cells are activated quickly—allowing the immune system to respond faster and more effectively than it did the first time. This is the principle behind vaccines: they introduce a weakened, inactivated, or fragmented version of a pathogen that is harmless but sufficient to stimulate an immune response and generate memory cells.

SYNTHESIS OF POLYAMINES: SPERMIDINE AND SPERMINE

Polyamines are small molecules that act as special nutrients within cells, helping to keep them healthy and functioning properly. Among these, spermidine and spermine stand out for their wide-ranging health benefits.

SAMe is not only involved in methylation; it can also be converted into another molecule called decarboxylated SAMe (dcSAMe). In this process, an enzyme removes a carboxyl group from SAMe, transforming it into dcSAMe, which acts as a donor of aminopropyl groups instead of methyl groups. These aminopropyl groups are essential for building the structure of polyamines by binding to a precursor molecule called putrescine.

Spermidine (named because it was first isolated from human semen) and spermine are two essential polyamines. The synthesis of both requires SAMe in its decarboxylated form (dcSAMe): first, to convert putrescine into spermidine, and then to convert spermidine into spermine.

Figure 1. Polyamine Synthesis

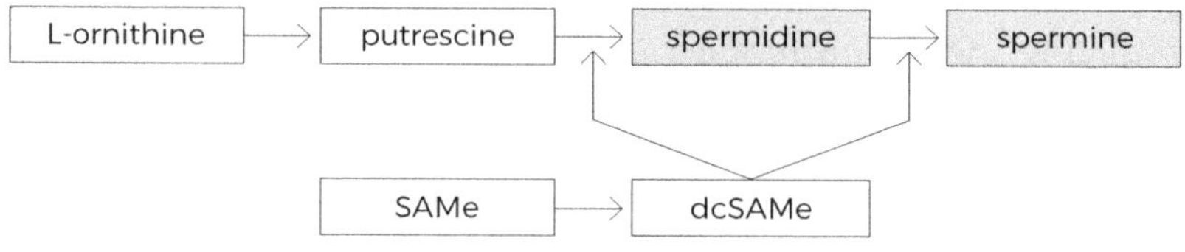

L-ornithine → putrescine → spermidine → spermine

SAMe → dcSAMe

Both spermidine and spermine have anti-inflammatory, antioxidant, cardioprotective, and neuroprotective effects. They also help extend lifespan, enhance mitochondrial biogenesis and efficiency, support musculoskeletal tissue regeneration, improve immune responses (both adaptive and cancer-specific), and inhibit tumor formation. However, their levels naturally decline with age.

Spermidine is particularly intriguing because, in addition to activating AMP-activated protein kinase (AMPK) and inhibiting the mammalian target of rapamycin (mTOR), it also inhibits the enzyme EP300. This enzyme regulates cell growth and division by acetylating proteins and activating genes that suppress autophagy. While this suppression is necessary during early life stages, promoting autophagy—the process of recycling damaged, aged, cancerous, or infected cells—is desirable in later life to slow aging and reduce age-related diseases. By inhibiting EP300, spermidine helps "release the brakes" on autophagy, enhancing it.

In addition to endogenous synthesis via SAMe, some gut microbiota bacteria can synthesize spermidine by digesting pectin—a type of fiber found in the white inner peel (albedo) of citrus fruits like oranges, lemons, and grapefruits. It's also found in notable amounts in the skin of green apples (such as Granny Smith) and fresh quince—not the processed versions commonly found in supermarkets.

In foods, spermidine is especially abundant in wheat germ, *nattō* (cooked and fermented soybeans), dried peas, shiitake mushrooms, and aged cheeses such as cheddar, Parmesan, and Manchego, as well as blue cheeses.

Although spermidine is also sold as a supplement (usually as wheat germ extract), in my opinion, it's not worth the expense. Just 20 grams of wheat germ or 25 grams of aged cheddar (aged at least one year) provide around 5 mg of spermidine—the maximum dose typically found in supplements.[20] Plus, getting it from food delivers additional beneficial nutrients.

A 2023 study[21] suggests that dietary spermidine may raise levels of its metabolite, spermine, more than spermidine itself. This would

---

[20] https://doi.org/10.3389/fnut.2019.00108
[21] https://doi.org/10.3390/nu15081852

imply that many of the benefits attributed to spermidine may actually be due, at least in part, to the downstream action of spermine. In any case, whether through endogenous synthesis or dietary intake, both pathways require adequate availability of SAMe—so it's essential to ensure sufficient levels.

Another way to support endogenous synthesis is through L-ornithine supplementation (at a dose of approximately 30 mg per kg of body weight). Since L-ornithine is a precursor to putrescine, which in turn leads to spermidine and spermine, its intake has been shown to boost levels of both polyamines.[22]

That said, recent studies have found that polyamines only extend lifespan and enhance autophagy when the enzyme GNMT is activated.[23] GNMT transfers a methyl group from SAMe to the amino acid glycine, converting it into sarcosine. The primary function of GNMT is to regulate the amount of SAMe available. Excess SAMe can cause problems, so GNMT helps maintain proper balance by "mopping up" surplus methyl groups and passing them to glycine, which acts like a "garbage truck." In other words, activating the GNMT pathway requires more SAMe than what's needed to fulfill all the other functions we've discussed.

In summary, for spermidine to exert its geroprotective effects and extend lifespan by enhancing autophagy, the following are required:

1. **Sufficient SAMe availability**: first, to enable the synthesis of spermidine; and second, to promote activation of the GNMT pathway, which is essential for spermidine to carry out its functions.
2. **Adequate glycine levels**: so that the GNMT enzyme can function properly.

Those who follow my blog or have read my previous book, *Evolutionary Nutrition*, already know this—but it's worth repeating: there is a widespread glycine deficiency in the general population of approx-

---

[22] https://doi.org/10.1016/j.bbrc.2019.03.147
[23] https://doi.org/10.1111/acel.13043

imately 10 grams per day.[24] The 4.5-6 grams of glycine we obtain through endogenous synthesis and diet is not enough. It's estimated that we need at least 16 grams per day for glycine to fully carry out its diverse and critical functions, which makes supplementation virtually "mandatory."

It has also been shown that simply overexpressing GNMT[25] and supplementing with glycine[26] can extend lifespan—even without spermidine. Not only is a deficiency in SAMe harmful, but an excess—caused by GNMT dysfunction or glycine deficiency—is also associated with certain diseases we'll explore in Chapter 6.

SYNTHESIS OF ANTIOXIDANTS: GLUTATHIONE AND TAURINE

Although indirectly, methylation and SAMe also play a key role in the synthesis of two of the body's most important antioxidants: glutathione at the intracellular level and taurine at the extracellular level.

As we discussed in Chapter 2, once SAMe donates its methyl group to another molecule, homocysteine is generated as a byproduct. We'll explore this in more detail in the next chapter, but one of the pathways for recycling homocysteine—known as the transsulfuration pathway—converts homocysteine into cysteine, which is the precursor for both antioxidants. This pathway is more active when SAMe levels are sufficient; otherwise, the body prioritizes remethylation pathways that convert homocysteine back into methionine to produce more SAMe.

In other words, the more SAMe we have available, the more we can ensure all methylation functions are fulfilled, which in turn generates more homocysteine as a byproduct. This enhances the transsulfuration pathway, producing more cysteine and enabling increased synthesis of antioxidants like glutathione and taurine. The resulting boost in antioxidant production has a highly positive impact on both health and longevity.

---

[24] https://doi.org/10.1007/s12038-009-0100-9
[25] https://doi.org/10.1038/ncomms9332
[26] https://doi.org/10.1016/j.arr.2023.101922

EYE HEALTH

SAMe plays a fundamental role in eye health by supporting the maintenance and repair of both the lens and the retina, as well as preventing issues in the eye's blood vessels.

The lens is a transparent structure located behind the iris and pupil that focuses incoming light onto the retina. The retina, in turn, is a layer of tissue at the back of the eye that contains light-sensitive cells called photoreceptors. When light—focused by the lens—reaches the retina, these photoreceptors convert it into electrical signals, which are then sent to the brain via the optic nerve. The brain interprets these signals as images. In other words, the lens and retina work together to capture and process light, allowing us to see clearly and in detail. If the lens fails to focus properly, the image on the retina becomes blurry, leading to common vision problems.

So, what role does SAMe play in all of this? Several, in fact:

- **Lens**: SAMe is found in high concentrations in the lens, where it helps methylate lens proteins—a process crucial for maintaining its structure and function.[27] As we age, SAMe levels in the lens decline. With less SAMe, the lens cannot properly maintain its proteins, leading to protein damage and dysfunction, which may result in cataracts or presbyopia.[28]
- **Retina**: Research has shown that high homocysteine levels and low SAMe availability are associated with problems in the small blood vessels of the retina—a condition known as retinal microangiopathy. These issues include retinopathy (damage to retinal blood vessels) and sclerosis (hardening) of retinal vessels.[29]

Moreover, as we've seen, SAMe is essential for synthesizing creatine, phosphatidylcholine, and glutathione—reactions that are vital for the function of retinal photoreceptors. When SAMe levels drop—as may

---

[27] https://doi.org/10.1016/j.exer.2012.04.002
[28] https://doi.org/10.1016/0014-4835(88)90003-6
[29] https://doi.org/10.1042/cs20070275

happen following retinal damage due to interrupted blood flow (retinal ischemia)—these SAMe-dependent reactions are impaired, leading to retinal dysfunction and compromised vision.[30] However, supplementation with SAMe after retinal injury has been shown to protect and restore photoreceptor function, improving ischemia-induced retinal damage and helping to restore vision.[31]

CONCLUSION

> **Without SAMe, there is no methylation—and without methylation, there is no life.**

We need the molecule SAMe and the process of methylation for all the functions we've discussed. Even just considering its role in DNA repair, genome stability, and gene expression through epigenetics, it becomes clear how crucial it is to have enough SAMe available to methylate everything we need—and ideally, a little extra to also support longevity pathways such as autophagy and antioxidant production.

When SAMe is insufficient, many of these functions become compromised. The body will always prioritize the most critical processes for immediate survival, pushing all others to the background.

In the practical section, we'll explore how nutritional strategies can help:

1. **Increase SAMe** to:
    a. Counter global hypomethylation and preserve genome stability.
    b. Silence genes that should remain inactive, enabling proper cellular differentiation.
    c. Support the many vital methylation-dependent functions beyond epigenetics that we've just explored.

---

[30] https://iovs.arvojournals.org/article.aspx?articleid=2363378
[31] https://doi.org/10.1017/S0952523809990241

2. **Inhibit the HDAC enzyme** (as we discussed in Chapter 1 —the enzyme that promotes gene silencing), which allows tumor suppressor genes and other important genes to remain active.

But we'll get there. To fully understand that part, we first need to learn how SAMe is "manufactured" (the methionine cycle), and what happens after methylation occurs (the homocysteine cycle).

KEY POINTS

1. **General methylation functions beyond epigenetics**:
    * SAMe donates methyl groups to molecules beyond DNA. Its role as a methyl group donor is not limited to epigenetics but also supports many other essential biochemical reactions throughout the body.

2. **Creatine synthesis**:
    * *In muscles:* Creatine is key for muscular energy, stored as phosphocreatine to improve strength, endurance, and recovery.
    * *In the brain:* It also provides energy to neurons, has antioxidant and neuroprotective effects, and promotes cellular development.

3. **Phosphatidylcholine synthesis**:
    * *Cell membranes:* Phosphatidylcholine is critical for cellular membrane integrity and fluidity, especially in the brain, where it protects against neurodegenerative diseases.
    * *Fat metabolism:* Helps prevent fat buildup in the liver and facilitates lipid transport.
    * *Choline synthesis:* Choline is a precursor to acetylcholine (a neurotransmitter vital for memory) and to trimethylglycine (which helps regulate homocysteine levels).

4. **Circadian rhythm regulation:**
   * *mRNA methylation:* SAMe is essential for regulating the body's biological clock. Disruption in methylation negatively affects functions such as sleep, metabolism, and hormone production.

5. **SAMe and the breakdown of catecholamines, estrogens, histamine, and niacinamide:**
   * *Catecholamines:* Helps break down neurotransmitters like dopamine and adrenaline via the COMT enzyme, preventing excess that could lead to stress dysregulation, cardiovascular issues, or mental health disorders.
   * *Estrogens:* Facilitates the excretion of catechol estrogens, supporting hormonal balance.
   * *Histamine:* Assists in degrading excess histamine, preventing exaggerated allergic reactions and migraines.
   * *Niacinamide (vitamin B3):* Converts excess niacinamide into excretable forms, preventing accumulation linked to neurological, metabolic issues, and certain cancers.

6. **SAMe and neurotransmitter/adrenaline synthesis:**
   * *Neurotransmitters:* Essential for the synthesis of serotonin, dopamine, and norepinephrine—regulators of mood, motivation, focus, and stress response.
   * *Adrenaline:* Involved in converting norepinephrine into adrenaline, crucial for the stress response.

7. **Immunity and T cells:**
   * *T cell differentiation:* SAMe is required for the epigenetic reprogramming of immune T cells, allowing their proliferation and specialization in defending the body.

8. **Polyamine synthesis:**
   * *Spermidine and spermine:* SAMe, in its decarboxylated form, participates in the production of polyamines essential for anti-inflammatory, antioxidant functions, and cellular longevity.

9. **Antioxidant synthesis:**
    - *Glutathione and taurine:* SAMe indirectly supports antioxidant production, which protects cells from oxidative damage and promotes longevity.

10. **Eye health:**
    - *Lens and retina:* SAMe is crucial for maintaining and repairing ocular structures, preventing issues like cataracts and retinopathy. Supplementation has been shown to support photoreceptor health.

11. **Overall conclusion:**
    - *The importance of SAMe:* Without it, methylation wouldn't occur—and methylation is vital for life and the proper functioning of numerous bodily systems. Maintaining adequate SAMe levels is essential for longevity and disease prevention, as it plays a role in a wide range of critical functions related to health and aging.

CHAPTER 4
Understanding the Methionine-Homocysteine Cycle

In the previous chapters, we explored the critical role of S-adenosylmethionine (SAMe) as a methyl group donor, both in epigenetic regulation and in other vital bodily functions. However, for these functions to take place efficiently and continuously, the body must maintain a constant supply of SAMe. This is where the methionine-homocysteine cycle comes into play—it ensures the ongoing production of SAMe and prevents the harmful accumulation of its byproduct, homocysteine.

Understanding this cycle is essential not only to grasp how SAMe is produced but also to see how its proper regulation influences aging, antioxidant balance, and disease prevention. In short, learning how this mechanism works will allow us to optimize SAMe availability and protect the body from the damaging effects of poor homocysteine management.

I've referred to it as a cycle, but it's actually more of a dual cycle or a two-phase cycle, as you'll see in the following diagrams. Each phase begins with two key molecules: methionine and homocysteine.

Let's start by giving it a bit of a dramatic introduction.

The methionine-homocysteine cycle is this complex process:

Figure 2. Methionine Cycle (Methylation Process)

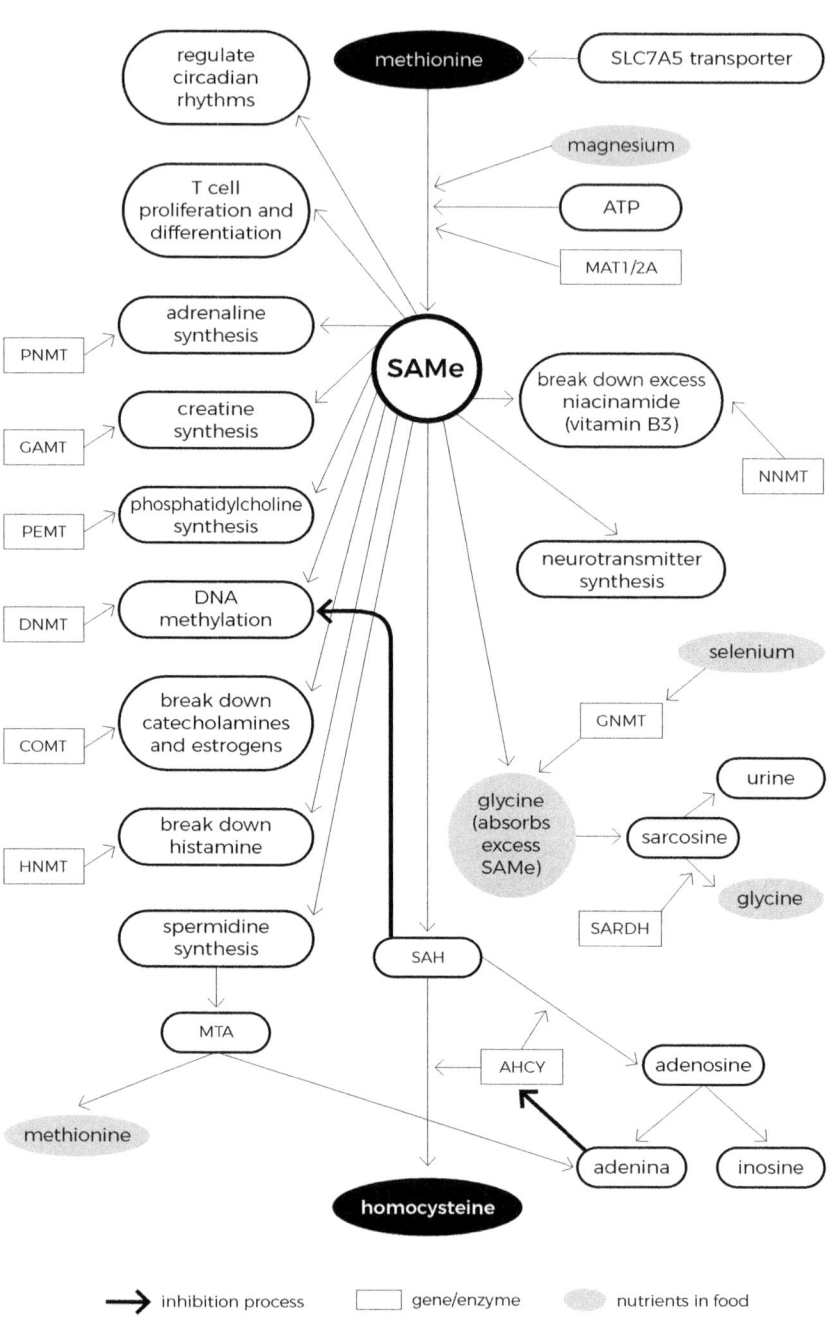

Figure 3a. Homocysteine Cycle (Remethylation Pathways)

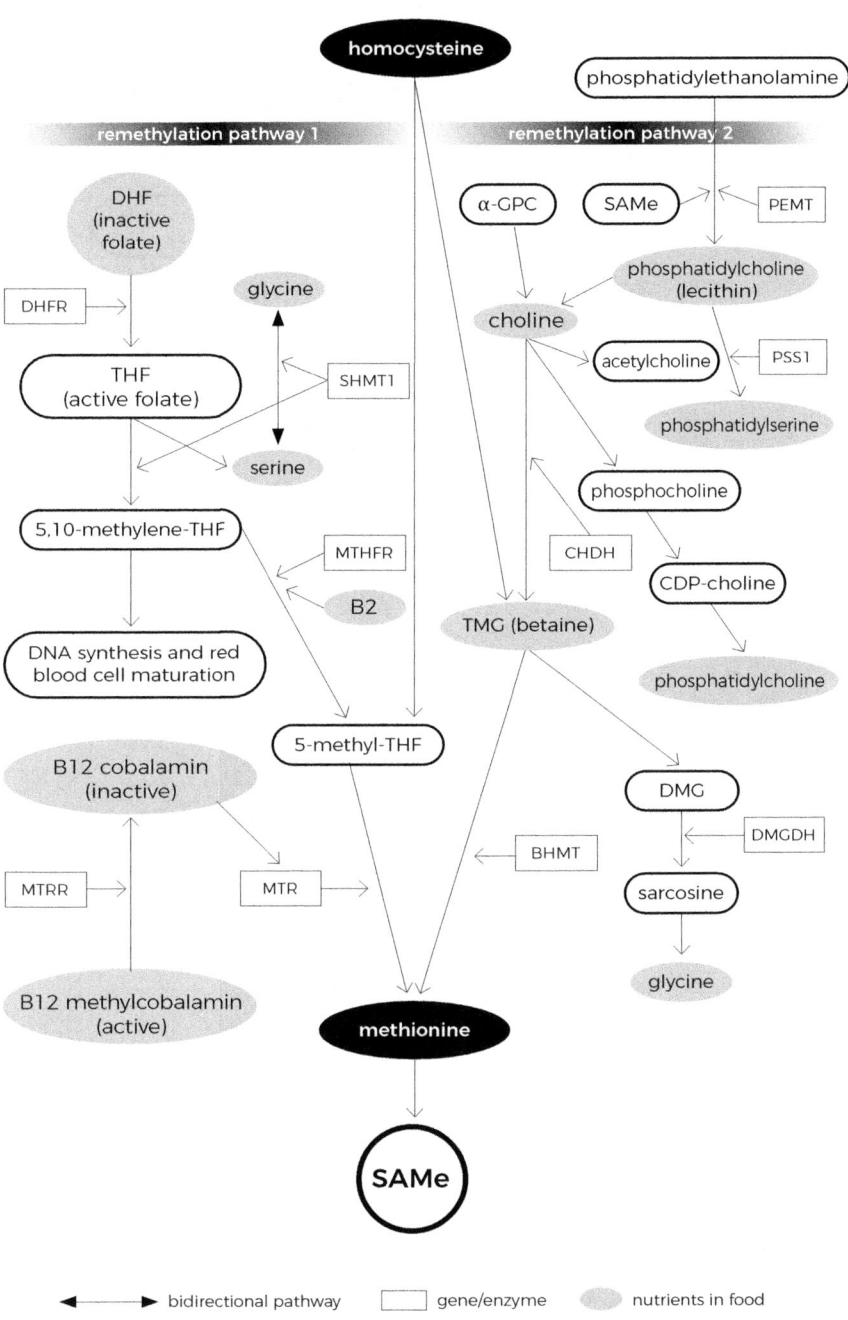

Figure 3b. Homocysteine Cycle (Transsulfuration Pathway)

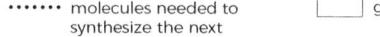

Feeling confused? That's perfectly normal—don't worry. Let's unravel this knot step by step.

THE METHIONINE CYCLE: HOW SAME IS PRODUCED

The goal of the first phase of the cycle is to synthesize SAMe and carry out the methylation processes. The cycle begins with methionine (see Fig. 2), an essential amino acid we get from our diet, mainly through animal-based proteins. When we consume foods containing methionine and metabolize them, a transporter known as SLC7A5 (L-type amino acid transporter 1 [LAT1]), located on the cell membrane, allows methionine to enter the cell.

Once inside the cell, methionine is processed by an enzyme called methionine adenosyltransferase (MAT), which converts it into SAMe. During this step, the methionine molecule combines with adenosine triphosphate (ATP), the cell's energy currency, to form our friend SAMe. Magnesium plays a key role here: it stabilizes the ATP, and without enough magnesium, this reaction could be disrupted.

After fulfilling its role as a methyl donor in all the processes we covered in the previous chapters, SAMe is converted into S-adenosylhomocysteine (SAH). At this stage, the molecule can no longer perform its function and needs to be recycled.

Through a series of steps, SAH is broken down into homocysteine. This is a critical juncture, because one of three things can happen to homocysteine: one bad, one good, and one even better.

The *bad* outcome is that nothing happens—it gets "stuck" as homocysteine. The *good* outcome is that it gets recycled into methionine (through remethylation), enabling the synthesis of more SAMe and restarting the cycle. The *best* outcome is that it gets converted into cysteine (through transsulfuration) and helps synthesize antioxidants. However, for this to happen, there must be a surplus of SAMe. If there isn't, the body will prioritize SAMe synthesis (for immediate survival) over antioxidant production (which benefits long-term health and slows aging).

Let's take a closer look.

THE HOMOCYSTEINE CYCLE: WHAT HAPPENS AFTER METHYLATION

Homocysteine is the inevitable byproduct of methylation—it's the molecule where the first phase of the cycle ends and the second begins. Once the first phase has completed its task (methylation), the second phase takes over to manage this leftover product, *recycling* homocysteine so it doesn't accumulate and can instead be reused in a beneficial way. To do this, the body relies on two different pathways (Figs. 3a and 3b), which I'll explain shortly.

We can think of the methionine-homocysteine cycle as a factory. Methionine is the raw material that enters the plant to produce vital products (like SAMe), while homocysteine is the waste that the factory must manage and recycle properly to prevent garbage buildup and ensure smooth, efficient operations.

As I mentioned earlier, homocysteine is the unavoidable byproduct of methylation. Every time the body carries out a methylation reaction, there's a decrease in SAMe and an increase in homocysteine, since the former is converted into the latter after donating its methyl group.

Homocysteine is an amino acid that isn't used to build proteins, and it's critical that it doesn't accumulate in the body because it's toxic. The purpose of this second phase of the cycle is precisely to eliminate homocysteine and prevent its buildup. Elevated levels of homocysteine in the blood (*hyperhomocysteinemia*) are linked to various diseases. In addition to damaging the inner lining of blood vessels (the endothelium)—which promotes plaque buildup and raises the risk of heart disease—it's also associated with neurological disorders like Alzheimer's and Parkinson's, as well as cancer, diabetes, migraines, mental health conditions, autoimmune diseases, and many others we'll explore in the next chapter.

Over millions of years, evolution has finely tuned the body to handle this problem of homocysteine accumulation in two elegant and brilliant ways—just as it always does.

Remethylation Pathways (Figure 3a)

In this first route, homocysteine is converted back into methionine, which can then be transformed into more SAMe and used to continue methylating. The body favors this pathway when SAMe levels are low—something that commonly occurs as we age, since, as we've seen, SAMe levels naturally decline over time. Because life depends on methylation, and because our ancestors didn't always have reliable access to protein-rich foods to obtain methionine and make SAMe, this pathway is vital for survival.

That's why the remethylation pathway is split into two sub-pathways, ensuring that if one doesn't function properly due to genetic factors or nutritional deficiencies, the other can help make up for it.

- **Remethylation Pathway 1 (Methionine Synthase Pathway [MTR])**: This sub-pathway depends on two vitamins. Active vitamin B9 (5-methyltetrahydrofolate [5-MTHF]) donates its methyl group to homocysteine, converting it back into methionine. In addition, active vitamin B12 (methylcobalamin) is required as a cofactor for the MTR enzyme that makes this process possible.
- **Remethylation Pathway 2 (Betaine-Homocysteine Methyltransferase Pathway [BHMT])**: Instead of vitamin B9, this sub-pathway uses betaine (also known as trimethylglycine [TMG]) to donate its methyl group to homocysteine, converting it into methionine

Transsulfuration Pathway (Figure 3b)

This pathway converts homocysteine into cysteine. Its main purpose is to produce glutathione, the body's most important intracellular antioxidant, and taurine, a versatile nonessential amino acid that, among other roles, also functions as an antioxidant—primarily in the extracellular space. This process depends on an enzyme called cystathionine beta-synthase (CBS), which requires vitamin B6 in its active form (pyridoxal-5-phosphate, or P-5-P) to function properly.

This pathway is always active to some degree, but it becomes more active when there is no shortage of SAMe—that is, when the body does not need to convert homocysteine back into SAMe. SAMe is always the top priority, because it's essential for survival. Naturally, the body prioritizes functions that support immediate survival over those that help slow aging, such as antioxidant production, which offer benefits in the medium to long term. After all, slowing aging is meaningless if we don't survive to experience it.

As we've seen, maintaining good SAMe availability not only ensures the many critical methylation functions we've discussed—improving the body's overall performance—but also enhances the transsulfuration pathway and promotes the production of antioxidants, which are essential for healthier, slower aging.

This second part of the cycle—the homocysteine cycle—is another remarkable example of the brilliance of our biology, refined by evolution to solve problems elegantly. Through this process, the body benefits in three key ways:

1. It avoids harm by breaking down homocysteine before it can accumulate and become toxic.
2. It recycles homocysteine to make more SAMe.
3. It converts homocysteine into antioxidants.

In short, this process eliminates a problem and creates two benefits. However, there's a catch: for everything to run smoothly, the body needs balance—and none of us is perfect. Due to genetic factors or nutritional deficiencies, this recycling process doesn't work flawlessly in everyone. Each of us has a weak link somewhere in the cycle. The severity and location of this weak point can be the underlying cause of a wide range of serious diseases, which we'll explore shortly. In the practical section of this book, we'll also learn how to fix—or at least minimize—the impact of these disruptions. Even in people without chronic conditions, there's room for improvement: we can optimize this cycle to reduce the effects of aging and improve our health and quality of life.

Before we get to that practical section, the next two chapters will examine which diseases are linked to excess homocysteine and SAMe deficiency (global hypomethylation), and—less common but also possible—SAMe excess (hypermethylation).

> **KEY POINTS**

1. Through the methionine cycle (Fig. 2), we produce the molecule SAMe, which is essential for carrying out all the body's vital methylation functions.

2. Once SAMe has completed its role, it is converted into homocysteine as a byproduct. It is crucial that homocysteine does not get "stuck" at this stage, as it can cause serious harm. Instead, it must continue through one of its two recycling pathways in the next phase of the cycle.

3. The goal of the homocysteine cycle is to prevent the accumulation of this toxic compound. It achieves this through three possible routes:
   - **Remethylation pathway** (Fig. 3a): Recycles homocysteine back into methionine to produce more SAMe when availability is low. Due to the importance of this process for survival, it includes two alternative routes:
     - **Sub-pathway 1**: Depends on the availability of active vitamins B9 and B12.
     - **Sub-pathway 2**: Relies on betaine, also known as trimethylglycine (TMG).
   - **Transsulfuration pathway** (Fig. 3b): Converts homocysteine into cysteine to synthesize two key antioxidants—glutathione and taurine. This pathway depends on active vitamin B6.

4. Having an ample supply of SAMe available for methylation is critical for all the functions we covered in Chapters 2 and 3. A deficiency in SAMe (hypomethylation) will negatively impact aging, accelerating overall deterioration and increasing the risk of disease. On the other hand, higher availability not only ensures that none of these essential functions are compromised but also activates two key pathways as we age: (1) the glycine N-methyltransferase (GNMT) pathway, which uses up excess SAMe and promotes autophagy, and (2) the transsulfuration pathway, which enhances antioxidant production.

CHAPTER 5
Diseases Linked to Global Hypomethylation

Before diving into the details, I want to start with a word of caution and two important clarifications to avoid any potential misunderstandings in this chapter.

Caution: Throughout this chapter, we'll encounter some technical and dense sections that may feel a bit challenging—or even boring—for some readers. If that's the case for you, I suggest focusing on the sections covering the diseases that interest you the most.

First clarification: Many of the epigenetic changes discussed here —because they appear in scientific literature in association with certain diseases—are correlational, not causal. It's crucial to remember that *correlation does not imply causation.* Just because a particular epigenetic change is associated with a disease doesn't necessarily mean it causes it. Experimentally proving that an epigenetic change triggers a disease is, to say the least, highly complex. This is important to keep in mind. Moreover, the interventions currently being developed to "correct" these epigenetic changes are still quite imprecise. Many of the epigenetic modifications proposed as therapeutic targets are not yet fully validated, which makes these approaches somewhat controversial among experts.

Second clarification: The methods used to detect epigenetic markers—such as "methylation clocks" and similar tools—are currently being questioned by experts due to inconsistencies in their results, especially those developed in the early stages. Sometimes, two tests using different procedures can yield very different outcomes, which makes drawing solid conclusions difficult. This variability may be due to the technique itself or to the diversity of cell types present in the analyzed samples, making it harder to measure these marks accurately—especially in studies on aging and age-related diseases. If epigenetics is to have a meaningful future as a field that can help solve health problems, this issue will need to be addressed and improved.

Now, let's move on. We've seen how the progressive decline in S-adenosylmethionine (SAMe)—caused by aging and by a malfunctioning methionine-homocysteine cycle, due to genetic predispositions or suboptimal nutrition—contributes to global DNA hypomethylation, epigenetic drift, and, consequently, a gradual and accelerated aging process. In the practical part of this book, we'll explore how to slow down this SAMe loss, but first let's look at how it relates to specific diseases.[32,33]

Let's begin with the ones that cause the highest number of deaths worldwide.

CARDIOVASCULAR DISEASES

Cardiovascular diseases (CVDs) encompass a wide range of disorders affecting the heart and blood vessels, including coronary artery disease (such as heart attacks or angina), cerebrovascular diseases (like stroke), high blood pressure, peripheral artery disease, rheumatic heart disease, congenital heart defects, and heart failure, among others. They are the leading cause of death worldwide and, in 2023, were responsible for approximately 20.5 million deaths. Yes, you read that right: that's the equivalent of half the population of a country like Spain—every single year.

The primary underlying mechanism behind many of these conditions is atherosclerosis, a process characterized by the buildup of fatty plaques (such as triglycerides and cholesterol carried by low-density lipoproteins, or LDL) on the inner walls of arteries (the endothelium), particularly when those walls are damaged and the fat particles are too small. This buildup can lead to narrowing and hardening of the arteries, reducing blood flow to the heart, brain, and other parts of the body.

---

[32] https://www.sciencedirect.com/topics/biochemistry-genetics-and-molecular-biology/dna-hypomethylation

[33] https://doi.org/10.1016/j.redox.2019.101322

Cholesterol

Before we go any further, I'd like to pause for a moment to talk about cholesterol, since it has been demonized to the point of being cast as the "villain" in cardiovascular disease—a message that has become deeply ingrained in the general population. In CVD, cholesterol is a necessary cause (because it's what accumulates), but it's not a sufficient cause—in other words, it's not the root of the problem. The common line of reasoning goes something like this: since cholesterol is what builds up to form the plaque that clogs blood vessels, let's eliminate it from the "equation."

In my view, this is both incorrect and overly simplistic. Cholesterol is our ally and plays vital roles in the body; without it, life wouldn't be possible. It's as if governments decided to ban all cars from the roads to prevent traffic accidents. Sure, accidents would stop—but we need cars to go about our daily lives. That approach would solve one problem while creating many others. No one would seriously propose eliminating all cars and allowing only the bare minimum to operate. Instead, we should look at why accidents happen in the first place —distracted or reckless drivers, poor road conditions, faulty vehicles— and focus on fixing those issues.

It's the same with cholesterol. It's not something that should be kept as low as possible; like everything in the body, it needs to be balanced and kept within an optimal range (which, according to studies,[34] is not the range we've been told to aim for). The key question is: why does cholesterol get lodged in the endothelium and build up, blocking blood vessels? That topic goes beyond the scope of this book, but I explored it in depth in my previous book, *Evolutionary Nutrition*, as well as in a dedicated article on my blog, for those who want to dive deeper. Let's continue.

The Silent Disease

Atherosclerosis is often called "the silent disease" because it doesn't present obvious symptoms. Depending on a person's lifestyle, this slow-developing condition can begin as early as adolescence, with the

---

[34] https://doi.org/10.1038/s41598-018-38461-y#Fig2

first warning sign being a serious event like a heart attack or stroke —sometimes fatal. Half of these events occur before the age of 65, and 25% before the age of 54.

Risk Factors

Risk factors for cardiovascular disease fall into two categories: non-modifiable and modifiable. Non-modifiable factors include age, sex, and genetic predisposition. But let's focus on the ones that matter most here—the modifiable ones. These are the lifestyle-related risk factors that directly influence epigenetics and play a significant role in the silent progression of the disease: smoking, physical inactivity, poor diet, excessive alcohol consumption, being overweight, nutritional deficiencies, and two often-overlooked causes—elevated homocysteine levels (due to a malfunctioning homocysteine cycle) and global DNA hypomethylation (caused by a SAMe deficiency).[35]

Hyperhomocysteinemia

Elevated homocysteine levels are a problem in and of themselves and a major contributor to cardiovascular disease (CVD). According to a meta-analysis[36] of 72 studies, lowering blood homocysteine by just 3 points could reduce the risk of ischemic heart disease by 16%, deep vein thrombosis by 25%, and stroke by 24%.

Homocysteine contributes to the loss of arterial wall elasticity by interfering with the relaxation mechanisms mediated by three key molecules:

- **Nitric oxide (NO)** derived from the endothelium, a vasodilator in large arteries.[37]
- **Endothelium-derived hyperpolarizing factor (EDHF)**, another vasodilator, but acting on small arteries.[38]

---

[35] https://doi.org/10.1373/49.8.1292
[36] https://doi.org/10.1136/bmj.325.7374.1202
[37] https://doi.org/10.1161/hc4601.098514
[38] https://doi.org/10.3390/ijms20153737

- **Elastin**, a protein that provides elasticity and resilience to arterial walls.[39]

This homocysteine-induced arterial stiffness also promotes high blood pressure.[40]

One of the main dangers of elevated homocysteine in the development of atherosclerosis is that it also fosters a state of global hypomethylation.[41] If homocysteine becomes "stuck" and is not remethylated back into SAMe (see Fig. 3a), there won't be enough methyl donors available to methylate DNA. Let's take a look at how this impacts the body.

Global hypomethylation

One of the main mechanisms by which homocysteine causes harm is through damage to the endothelium—the inner lining of blood vessels. As we've already seen, when the endothelium is damaged, fat particles embed themselves into its walls and begin to form atherosclerotic plaque, narrowing the vessels and raising cardiovascular risk.

Homocysteine accelerates the aging of endothelial cells by inactivating telomerase, the enzyme responsible for keeping telomeres healthy. Telomeres protect DNA and its genetic information and are essential for cellular health and genomic stability. By promoting hypomethylation, homocysteine reduces the expression and activity of human telomerase reverse transcriptase (hTERT), a key component required for telomerase to function properly. The good news is that this effect can be reversed by increasing SAMe levels.[42]

Another negative consequence of high homocysteine-induced hypomethylation for endothelial health is its ability to slow cell growth by inhibiting the transcription of the cyclin A gene.[43,44] Cyclins are essential proteins that regulate the cell cycle. Before new cells can form, DNA must be duplicated so that two full sets of chromosomes can be

---

[39] https://doi.org/10.1016/j.amjhyper.2003.12.002
[40] https://doi.org/10.1161/ATVBAHA.116.308515
[41] https://doi.org/10.15406/jccr.2023.16.00575
[42] https://doi.org/10.1161/ATVBAHA.114.303899
[43] https://doi.org/10.1182/blood-2007-06-096701
[44] https://doi.org/10.1007/s11239-011-0550-4

passed on during cell division. Cyclin A plays a critical role in this stage, so when it's inhibited, endothelial cell growth in the "intimal layer"—the inner layer that acts as a barrier between the bloodstream and the other layers of the vessel wall—is reduced.

It's not just excess homocysteine that leads to hypomethylation in cardiovascular disease. Studies have shown that, even independently of homocysteine, global methylation levels tend to decline with age as SAMe levels drop. This global hypomethylation causes structural changes and instability in chromosomes.[45] According to a study[46] from the University of Finland, reversing global DNA hypomethylation is a highly promising therapeutic strategy for atherosclerosis—and it might be as simple as increasing circulating levels of SAMe.

CANCER

To describe cancer cells in just a few words, I can't think of a better definition than the one Carlos López-Otín uses in the title of his book, *Egoístas, inmortales y viajeras* (*Selfish, Immortal, and Nomadic*). Cancer cells are *selfish* because they defy programmed cell death, prioritizing their own survival over the well-being of neighboring cells, future cells, and even the organism to which they belong. They are *immortal* because they find ways to evade the mechanisms that would normally lead them to die. And they are *nomadic* because they spread relentlessly, invading other areas of the body in search of the resources they need to survive.

Prevention

Although science has made tremendous progress in recent decades —both in early detection and in treatment—and many malignant tumors are now curable in a growing percentage of cases, prevention is still by far the smartest strategy. Fortunately, there's an increasing, though still insufficient, awareness of the importance of:

---

[45] https://www.frontiersin.org/journals/pharmacology/articles/10.3389/fphar.2022.815977/full#h3

[46] https://doi.org/10.1093/eurheartj/ehu437

1. **Nutrition**: not just *what* we eat, but—more importantly—*what we don't eat, how much* we eat, and *when* we eat.
2. **Physical activity**: which doesn't necessarily mean playing sports, but simply moving enough throughout the day to avoid a sedentary lifestyle.
3. **Stress reduction and management**: one of the great afflictions of modern life that, indirectly, raises the risk of nearly every chronic illness.
4. **Avoiding exposure to toxins**: including tobacco, alcohol, air pollution, and carcinogenic substances in everyday products.

Now we'll explore a fifth area of cancer prevention—one that's equally important but often overlooked: slowing the age-related decline of SAMe levels in order to enhance global DNA methylation capacity.

Increasing SAMe and Methylation Capacity

As early as 1983, researchers detected significantly lower levels of 5-methylcytosine (5mC) in tumor tissues compared to normal tissues.[47] Let's recall that 5mC is the molecule formed when SAMe donates its methyl group and is used by the body to methylate DNA, helping to determine when a gene should remain silent. Although this discovery didn't receive much attention at the time, it was one of the first signs of global DNA hypomethylation associated with cancer. Despite the initial findings and the growing number of studies confirming the phenomenon, the subject continued to be largely overlooked. In 2010, a study by the Department of Biochemistry and Cancer Center at the School of Medicine in New Orleans,[48] lamented this very fact, concluding that "DNA hypomethylation in cancer can no longer be regarded as an oddity, because recent high-resolution whole-genome studies confirm that DNA hypomethylation is an almost constant companion to genome-wide hypermethylation in cancer […] and that evidence is mounting for the biological significance and clinical relevance of DNA hypomethylation in cancer."

---

[47] https://doi.org/10.1093/nar/11.19.6883
[48] https://doi.org/10.2217/epi.09.33

Today, although still not widely known outside the scientific community, the decline in SAMe—and consequently, the loss of DNA methylation—in cancer is an undeniable fact, backed by numerous studies. To cite just one recent example: a 2023 study from the University Medical Center Utrecht[49] found that carcinogenesis is accompanied by a global loss of DNA methylation, which facilitates cellular transformation. According to the study's authors, the main consequence of this hypomethylation in cancer is genomic instability—due to interference with the protective function of telomeres, which, as we saw, also plays a role in cardiovascular disease—and the acceleration of cancer progression. Moreover, the greater the loss of methylation, the higher the risk of metastasis.[50]

Although this has been confirmed by many studies, even without them it makes complete biological sense: as discussed in the previous chapter, methylation is fundamental to proper gene expression, contributing to genomic integrity and stability, and protecting against the genetic mutations that underlie cancer.

Do you remember that we also discussed in Chapter 2 how global DNA hypomethylation leads to hypermethylation (reduced expression) of specific genes, including tumor suppressors? A 2023 study[51] by the Cancer Center at Massachusetts General Hospital (Harvard Medical School) confirms that the loss of DNA methylation also results in the silencing of antitumor immune genes. In other words, it not only promotes the onset of the disease but later weakens the body's innate mechanisms to fight it.

Given this, it makes sense that a first step would be to evaluate methylation capacity (as we'll see in the first chapter of the practical section), since, as several studies have shown, it's a useful epigenetic biomarker for early detection and cancer risk assessment.[52,53]

---

[49] https://doi.org/10.1038/s41598-023-33932-3
[50] https://pubmed.ncbi.nlm.nih.gov/24294382/
[51] https://doi.org/10.1016/j.cell.2023.05.028
[52] https://doi.org/10.1158/1055-9965.epi-12-0859
[53] https://doi.org/10.1186/s12885-015-1461-0

Once methylation capacity is known, if it's confirmed to be insufficient—or even if it's adequate, but we want to prevent its decline with age—a logical next step would be to increase the amount of available SAMe in the body (which we'll also cover in the practical section). As highlighted in a 2022 study[54] by the Division of Experimental Pathology and Oncology at the University of Sassari (Italy), SAMe inhibits the growth of precancerous and cancerous cells (liver, breast, head, neck, pancreas, prostate, colorectal, melanoma, among others) and increases genome stability, making it an effective therapeutic agent both on its own and in combination with current pharmacological treatments.

NEURODEGENERATIVE DISEASES

Neurodegenerative diseases are a group of disorders characterized by the progressive degeneration of the structure and function of nerve cells—or neurons—in the brain and spinal cord. This gradual deterioration leads to the loss of cognitive abilities, such as memory and judgment, as well as motor skills like walking and speaking, along with other nervous system functions. In this section, we'll explore the most common neurodegenerative diseases and their connection to reduced methylation capacity.

Alzheimer's Disease

Unfortunately, we're all familiar with the hallmarks of this devastating and increasingly prevalent disease: the loss of memory and mental capacity. Unlike cancer, attempts to cure Alzheimer's disease have failed repeatedly. In my humble opinion, this is because they focus on treating the symptoms or consequences of the disease instead of addressing its root causes. Therefore, while prevention is always crucial, in the case of Alzheimer's, prevention becomes even more essential.

---

[54] https://doi.org/10.3390/cells11030409

As early as 2011, a study[55] involving both healthy individuals and Alzheimer's patients showed that the group with worse mental status results had lower levels of methylation compared to those with better results. Just as with cancer, studies continued to accumulate confirming the low global methylation capacity in people with Alzheimer's (and the correlation between lower methylation and increased dementia), reinforcing the early evidence. However, as with cancer, this finding also failed to attract much attention.

The Beta-Amyloid Plaque Hypothesis

Since the 1990s, resources for Alzheimer's treatment have been primarily directed toward drugs aimed at reducing the well-known beta-amyloid plaques found in many patients with this condition. Despite the enormous investment, only a handful of extremely expensive drugs have been developed that successfully reduce plaque levels—but these have yielded minimal improvements in slowing disease progression.

Beta-amyloid peptides are found both in the brain and in the bloodstream and serve several fundamental biological functions: they enhance neuronal signaling and communication (increasing synaptic strength, which is essential for learning and memory); stimulate neuronal differentiation; support brain injury recovery; inhibit oxidative stress; possess antimicrobial activity; and suppress the growth of tumor cells.[56]

So, much like cholesterol in cardiovascular disease, the issue is not the beta-amyloid peptide itself but rather its accumulation into plaques, which become neurotoxic and cause brain damage. As with any effective problem-solving process, the first step is identifying the root cause. Therefore, the key question is: what triggers beta-amyloid to accumulate into plaques, becoming neurotoxic and damaging the brain?

Alternative Hypotheses

For some time now, certain researchers have argued that beta-amyloid plaque is not the cause of Alzheimer's disease but rather a con-

---

[55] https://doi.org/10.1016/j.bbi.2011.01.017
[56] https://doi.org/10.1038/s41392-023-01484-7

sequence.[57,58] Gradually, this has opened the door to other lines of investigation:

- **Acetylcholine deficiency**: Lower-than-normal levels of the neurotransmitter acetylcholine have been detected in the brains of individuals with Alzheimer's. Acetylcholine is critical for attention, memory, and learning. In fact, one of the most commonly prescribed drugs for mild to moderate Alzheimer's is cholinesterase inhibitors—these inhibit the enzyme responsible for breaking down acetylcholine in the nervous system. By blocking this enzyme, the concentration and activity duration of acetylcholine in neuronal synapses are increased. However, as is often the case, this treatment addresses the symptom rather than the root cause. Again, the important question here is: what causes the low levels of acetylcholine?

    Acetylcholine is synthesized from the nutrient choline, which we can obtain from the diet, but it can also be synthesized endogenously from phosphatidylcholine using SAMe. A deficiency of choline—whether due to a poor diet, low SAMe levels, or both—could be the underlying cause of reduced acetylcholine. Moreover, choline is also necessary for other functions, such as being metabolized into trimethylglycine (TMG), also known as betaine, which is needed to eliminate homocysteine. Therefore, elevated homocysteine could contribute to acetylcholine deficiency, as choline would be diverted to produce TMG, reducing its availability for neurotransmitter synthesis.

    Thus, we can identify two methylation-related pathways potentially involved in the acetylcholine deficiency issue:
    1. SAMe deficiency, and
    2. an inefficient homocysteine cycle.

- **Damage to the Blood-Brain Barrier**: Another well-documented finding in Alzheimer's patients is the presence of a compromised

---

[57] https://doi.org/10.1016/j.jalz.2013.11.003
[58] https://doi.org/10.1038/nn.4017

blood-brain barrier.[59] This barrier is made up of endothelial cells lining the brain's blood vessels and serves to regulate the passage of substances from the bloodstream into the brain. Its primary role is to protect the brain by allowing essential nutrients to pass through, blocking potentially harmful substances (such as toxins and pathogens), and removing beta-amyloid to prevent its accumulation. In other words, when the barrier is damaged, it not only facilitates the buildup of beta-amyloid plaques but also allows the entry of harmful agents. Both of these outcomes can trigger overactivation of microglia—the brain's resident macrophages (a type of white blood cell)—which help eliminate beta-amyloid and attack invading pathogens through inflammatory responses. This excessive neuroinflammation causes neuronal damage, impairs synaptic plasticity and learning, and ultimately triggers the symptoms of Alzheimer's. Furthermore, the inflammation also harms endothelial cells, degrading blood vessels and weakening the blood-brain barrier even more—creating a self-perpetuating vicious cycle.

This leads us back to our favorite kind of question: *What's damaging the barrier?*

Several factors are at play. For one, there are genetic predispositions such as the APOE4 allele, which slows the clearance of beta-amyloid by a factor of two to three.[60] Additionally, studies have shown reduced levels of the glucose transporter GLUT1 in the blood-brain barrier, leading to microvascular degeneration in the brain, reduced and dysregulated cerebral blood flow, degradation of the barrier, and thus further beta-amyloid accumulation.[61]

Insulin resistance and type 2 diabetes both exacerbate this condition by impairing glucose transport and promoting neuroinflammation, thereby damaging the blood-brain barrier and accelerating disease progression.

---

[59] https://www.frontiersin.org/journals/aging-neuroscience/articles/10.3389/fnagi.2023.1102809/full#B139

[60] https://doi.org/10.1172/jci36663

[61] https://doi.org/10.1038/nn.3966

Moreover, as we saw in the section on cardiovascular disease, elevated homocysteine levels also damage the endothelium —meaning that high homocysteine levels contribute to the breakdown of the blood-brain barrier as well. Let's take a closer look at homocysteine.

Hyperhomocysteinemia

There are at least seven ways in which elevated homocysteine levels contribute to the pathogenesis of Alzheimer's disease, making it a significant—yet fortunately modifiable—risk factor for cognitive decline and dementia.[62] We've already discussed two of these mechanisms:

1. **Reduction of the neurotransmitter acetylcholine**: Homocysteine reduces the availability of choline through two main pathways:
   a. By consuming choline to synthesize trimethylglycine (TMG), which is needed to lower homocysteine levels.
   b. By decreasing SAMe, which is essential for choline synthesis.

2. **Breakdown of the blood-brain barrier**: Homocysteine damages the endothelium in four ways:
   a. By deactivating telomerase.
   b. By inhibiting cyclin A (both mechanisms already discussed in the cardiovascular disease section).
   c. By increasing levels of asymmetric dimethylarginine (ADMA), which inhibits the enzyme responsible for nitric oxide (NO) production.
   d. By elevating superoxide levels, a precursor of reactive oxygen species (ROS), which in turn intensifies oxidative stress.[63]

The other five mechanisms are as follows:

---

[62] https://doi.org/10.3233/JAD-171042
[63] https://doi.org/10.1161/01.STR.0000115161.10646.67#FIG1

1. **Contribution to the formation of "tau neurofibrillary tangles"**: Tau is a protein found in neurons that binds to microtubules—tiny "train tracks" that transport nutrients and messages throughout the brain. Problems arise when tau becomes overactivated (technically, hyperphosphorylated), causing it to twist into tangles inside neurons. This disrupts microtubule stability, interrupting internal brain communication and leading to neuronal dysfunction and cell death, which deteriorates brain function. Homocysteine contributes to the formation of tau tangles in three ways:
    a. By activating the enzyme Cdk5, which phosphorylates (activates) tau.[64]
    b. By inhibiting protein phosphatase 2A (PP2A), which dephosphorylates (deactivates) tau.[65]
    c. By activating caspase-3, which, when overactive, cleaves tau into smaller fragments that more readily misfold and clump into tangles.[66]

2. **Causing hypertrophy (thickening) of small cerebral blood vessels**: This includes both cellular and structural damage to these vessels, such as the deterioration of the endothelial cells lining the vessels and the pericytes surrounding them, destruction of mitochondria (the cell's energy generators), and cytoplasmic breakdown (a critical component of cell function and survival). Thickening of the vessel walls also reduces their ability to dilate, which impairs cerebral blood flow regulation.[67]

3. Homocysteine overactivates NMDA receptors—glutamate neurotransmitter receptors—[68] which triggers oxidative stress, neuroinflammation, beta-amyloid production and deposition,

---

[64] https://doi.org/10.1002/ana.24145
[65] https://doi.org/10.1016/j.neurobiolaging.2007.04.015
[66] https://doi.org/10.3390/ijms19030891
[67] https://doi.org/10.1093/jn/132.11.3418
[68] https://doi.org/10.3390/ijms22126259

synaptic dysfunction, and neuronal death.[69] In fact, memantine, a drug used in later stages of the disease, is an NMDA receptor inhibitor.

4. **Promoting beta-amyloid accumulation and plaque formation in neurons**: This occurs via two processes:
   a. Destabilization of the amyloid precursor protein (APP),[70] causing it to misfold and break into smaller fragments such as beta-amyloid.
   b. Reduction of apolipoprotein E,[71] which not only transports fats for use by cells—including neurons—but also plays a role in clearing beta-amyloid from the brain. Furthermore, a 2024 study[72] confirmed that reduced ApoE not only contributes to elevated beta-amyloid levels but also increases its toxicity, thereby worsening the disease. This might explain why some individuals with high beta-amyloid levels remain symptom-free, while others develop symptoms despite lower levels.

5. **Reduced availability of SAMe, leading to decreased methylation capacity**: As we've seen, if homocysteine builds up and is not properly recycled, it cannot be used to produce antioxidants (see Fig. 3b) or to generate more SAMe (see Fig. 3a), impairing the methylation process. For instance, vitamin B12 —which is essential for converting homocysteine back into SAMe (as discussed in Chapter 9)—has been shown to protect against Alzheimer's by mitigating beta-amyloid effects, lowering homocysteine levels, and boosting SAMe availability.[73]

---

[69] https://doi.org/10.1007/s10072-016-2546-5
[70] https://doi.org/10.2353/ajpath.2009.081036
[71] https://doi.org/10.3233/JAD-141101
[72] https://doi.org/10.1111/acel.14106
[73] https://www.sciencedirect.com/science/article/pii/S2211124721012079#sec3

SAMe Deficiency and Global DNA Hypomethylation

As we mentioned earlier, in 2011 global hypomethylation began to be detected in Alzheimer's patients, and it was observed that the lower the level of methylation, the more severe the symptoms of the disease. While the "official" approach remained focused on beta-amyloid plaques, studies confirming this hypomethylation continued to accumulate. In 2019, the Stockholm Center for Alzheimer's Research and Aging, the Department of Neurology at Ulm Hospital (Germany), the Nutrition Institute at the University of Oslo, and the Department of Pharmacology at the University of Oxford published the results of a six-year study[74] involving 2,570 individuals aged 60 to 102 who had no prior diagnosis of dementia. The results showed that higher values in the methionine-to-homocysteine ratio (more methionine and less homocysteine—a proxy for methylation capacity, as we'll explore in the practical section) were significantly associated with a lower risk of dementia and Alzheimer's, as well as with a slower rate of total brain tissue volume loss. The study concluded that a higher methionine-to-homocysteine ratio may play an important role in reducing brain atrophy and lowering the risk of dementia in older adults.

In 2022, a follow-up study[75] of a Japanese population over the age of 65 found that lower SAMe levels were significantly associated with a higher risk of both dementia and death. Also in 2022, another study[76] pointed out that—given the failure of current drugs due to their low efficacy and high incidence of side effects—epigenetics holds great potential as a new therapeutic target for treating Alzheimer's. Then, in 2023, yet another study[77] concluded: "SAMe depletion contributes to multiple pathophysiological processes observed in Alzheimer's brains, including the silencing of genes involved in disease development and progression, acetylcholine depletion, and impaired tau dephosphorylation."

---

[74] https://doi.org/10.1001/jamapsychiatry.2019.1694
[75] https://doi.org/10.1038/s41598-022-16242-y#Sec10
[76] https://doi.org/10.3389/fnagi.2022.911635
[77] https://link.springer.com/article/10.14283/jpad.2023.55#Sec29

I suppose some people may still want more evidence, but as far as I'm concerned, this is more than enough to justify efforts to lower homocysteine and increase SAMe levels.

In addition to everything mentioned, a deficiency in SAMe negatively affects the body's ability to synthesize phosphatidylcholine (see Figs. 1 and 2a). As we discussed in the chapter on the functions of methylation, SAMe is essential for synthesizing phosphatidylcholine, a key component of cell membranes—including neuronal membranes—since it supports both their structural integrity and fluidity (which is vital for communication), while also protecting them from oxidative damage. Another crucial role of phosphatidylcholine in brain health is its ability to facilitate the transport of omega-3 docosahexaenoic acid (DHA) into the brain and its incorporation into neuronal membranes. DHA is another critical component of neuronal membranes and not only possesses anti-inflammatory and anti-oxidant properties—both of which are linked to the development of neurodegenerative diseases—but also supports neurogenesis (the creation of new neurons). For all these reasons, DHA is essential for maintaining brain function, including memory, learning, and cognitive capacity.[78]

Parkinson's Disease

After Alzheimer's, Parkinson's disease is the second most common neurodegenerative condition among older adults and the fastest growing. Its incidence doubled between 1990 and 2015, and it's projected to double again by 2040. Parkinson's is a progressive neurodegenerative disorder that affects movement. It's caused by the death of nerve cells in a part of the brain called the *substantia nigra*, leading to reduced levels of dopamine—a neurotransmitter critical for movement control and coordination.

The main symptoms include resting tremors, muscle stiffness, slowness of movement (bradykinesia), and balance issues. Non-motor symptoms may also occur, such as sleep disturbances, hallucinations, cognitive problems, depression, or anxiety. These non-motor symp-

---

[78] https://doi.org/10.3390/nu13030986

toms appear to be more closely linked to the presence of *Lewy bodies* —abnormal intracellular inclusions primarily composed of the neuronal protein alpha-synuclein, found in the neurons of people with Parkinson's and other neurodegenerative conditions known as *Lewy body dementias*.

The exact cause of Parkinson's remains unknown. However, in addition to potential genetic and environmental factors (such as diet or exposure to heavy metals or pesticides), epigenetic factors—which involve the interaction between genes and environment—play a key role. Although there's currently no cure, improving epigenetic function by enhancing methylation capacity through increased SAMe levels may help prevent the disease, improve the effectiveness of standard drug treatments while reducing their side effects, and provide better symptom management. Let's talk about the medication.

Pharmacological Treatment and Its Side Effects (Dyskinesia)

Levodopa (L-DOPA) remains the most commonly used medication in the treatment of Parkinson's disease. Its mechanism of action is to increase dopamine levels, as L-DOPA is a precursor of this neurotransmitter. The problem is that, over time and as the disease progresses, the medication becomes less effective, requiring higher doses to manage symptoms. This increase can lead to side effects, the most common of which is dyskinesia—uncontrollable, erratic, writhing movements of the face, arms, legs, or torso.

It has been shown[79] that L-DOPA induces the overexpression of vascular endothelial growth factor (VEGF), a powerful mediator of angiogenesis—the formation of new blood vessels from existing ones. While angiogenesis is essential for tissue repair and wound healing by improving oxygen and nutrient supply to damaged tissues, its dysregulation can contribute to various pathologies such as cancer and macular degeneration in the retina.

In Parkinson's disease, L-DOPA administration leads to large, intermittent increases in extracellular dopamine levels, which in turn trigger VEGF overexpression and, consequently, angiogenesis. This

---

[79] https://doi.org/10.1093/brain/awr165

process causes exaggerated increases in cerebral blood flow, resulting in damage (increased permeability and inflammation) to the blood-brain barrier and leading to dyskinesia.[80]

Additionally, a compromised blood-brain barrier allows more pro-inflammatory molecules, immune cells, and other potentially harmful substances to enter the brain, promoting neuroinflammation and neuronal damage—features common to all neurodegenerative diseases. Subsequent studies[81] have also shown that, beyond damaging the blood-brain barrier, angiogenesis is associated with gait difficulties and more pronounced brain lesions (in the white matter and microhemorrhages).

These factors may partly explain the long-term decline in the drug's effectiveness and the worsening of disease symptoms over time. Now let's look at how SAMe is related to the medication.

Need for SAMe in Pharmacological Treatment

L-DOPA requires SAMe in order to synthesize dopamine, as SAMe provides the methyl groups necessary for the methylation reactions involved in dopamine production. In other words, without SAMe, there is no dopamine—regardless of how much of the drug is taken. Thus, when SAMe levels are low, the drug loses effectiveness in fulfilling its primary function. On the other hand, because L-DOPA depends on SAMe to work, it depletes SAMe,[82] diverting it away from its other essential functions.

By depleting SAMe, L-DOPA inactivates the enzyme PP2A,[83] which not only functions as a tumor suppressor but is also responsible for deactivating the tau protein. This promotes the formation of tau tangles, which contribute to neuronal dysfunction and cell death—particularly in cases of active folate (vitamin B9) deficiency (a topic that will be covered in the practical section). The intracellular accumulation of tau protein was underestimated for some time in Parkinson's disease,

---

[80] https://doi.org/10.1016/j.neuroimage.2012.02.066
[81] https://doi.org/10.1212/WNL.0000000000002151
[82] https://doi.org/10.1016/s0024-3205(00)00557-9
[83] https://doi.org/10.1523/JNEUROSCI.0125-12.2012

but it is now known to be one of the major contributors to the disease's pathogenesis.[84]

Additionally, because L-DOPA requires SAMe for both dopamine synthesis and for the degradation of excess dopamine through the enzyme catechol-O-methyltransferase (COMT), it raises homocysteine levels[85, 86]—with all the harmful consequences we have already discussed.

Hypomethylation as a Risk Factor

As if it weren't enough that the medication significantly depletes SAMe levels, patients with Parkinson's disease already have lower levels of SAMe than healthy individuals—nearly half,[87] according to some studies. This results in a substantial reduction in the global DNA methylation capacity, with decreases of up to 30% reported.[88]

Among other consequences, this leads to hypomethylation of the *SNCA* gene, which encodes the previously mentioned alpha-synuclein protein. When this gene is less methylated (and thus less silenced) than normal, it becomes overexpressed. This epigenetic dysregulation leads to two main problems:

1. **Formation of Lewy bodies** due to the overproduction of alpha-synuclein, which contributes to neuronal deterioration and cell death.
2. **Reduction of DNA methyltransferase (DNMT) enzymes**, which require SAMe to methylate DNA—further lowering overall methylation levels and triggering a vicious cycle.[89]

Raising SAMe levels not only improves the reduced DNA methylation capacity, but has also been shown to:

---

[84] https://doi.org/10.1016/j.biocel.2010.07.016
[85] https://doi.org/10.1016/j.parkreldis.2023.105357
[86] https://doi.org/10.1002/mds.20261
[87] https://doi.org/10.1006/exnr.1997.6466
[88] https://doi.org/10.1074/jbc.C110.212589
[89] https://doi.org/10.1111/jnc.13646

1. **Protect dopaminergic neurons** against L-DOPA-induced toxicity[90] by promoting the breakdown of its excess through the COMT enzyme (see Fig. 2).
2. **Inhibit VEGF**, and therefore, the angiogenesis induced by L-DOPA.[91]
3. **Improve depression in Parkinson's patients**[92] at doses ranging from 800 to 3,600 mg/day.

However, it's important to note that excessively high doses of SAMe may lower dopamine levels[93,94] and potentially mimic Parkinsonian symptoms—though this effect only occurs at extremely high intakes (we will address precautions regarding SAMe supplementation in the practical section). In contrast, at normal doses (no more than 48 mg/kg of body weight), a 2022 study[95] shows that SAMe:

- Improves motor coordination,
- Reduces oxidative stress by increasing glutathione,
- Prevents dopaminergic neuronal loss,
- Restores dopamine levels, and
- Enhances the methylation status of DNA in dopaminergic neurons.

Exercise

Although it goes beyond the scope of this book, it's important to highlight the critical role of exercise in both the prevention and improvement of symptoms in virtually any disease—neurodegenerative or otherwise. In the case of Parkinson's disease, a 2022 systematic review and meta-analysis conducted by Beijing Sport University[96] found that regular aerobic exercise improves balance, gait (including speed and stride length), and motor function in patients. It also sig-

---

[90] https://doi.org/10.1016/S0006-8993(00)03298-4
[91] https://doi.org/10.18632/aging.103863
[92] https://pubmed.ncbi.nlm.nih.gov/11104210/
[93] https://doi.org/10.1016/0163-1047(93)90950-M
[94] https://doi.org/10.1016/j.lfs.2008.09.020
[95] https://doi.org/10.1002/ar.24948
[96] https://doi.org/10.1038/s41531-022-00418-4

nificantly reduces the risk of developing the disease in those who don't already have it.

In the next sections on other neurodegenerative diseases, I won't go into as much detail, since the underlying mechanisms are largely the same as those we've already covered in relation to Alzheimer's and Parkinson's.

Amyotrophic Lateral Sclerosis

Amyotrophic lateral sclerosis (ALS) is a progressive and fatal neurological disorder that affects nerve cells in the brain and spinal cord. The disease leads to the degeneration and death of motor neurons, resulting in the loss of voluntary muscle control. Early symptoms include muscle weakness and difficulty speaking, swallowing, and eventually breathing. As with other neurodegenerative diseases, the cause of ALS is likely a combination of genetic and environmental factors, and currently, there is no cure.

As early as 2009, researchers began observing that SAMe supplementation could delay the onset of ALS and help mitigate features of neurodegeneration, including the prevention of motor neuron loss and the activation of antioxidant responses.[97] These findings align with the lower levels of methionine (the precursor to SAMe) detected in people with ALS compared to healthy individuals,[98] along with reduced levels of cysteine and glycine—both of which are precursors to the antioxidant glutathione. Another hallmark of ALS is the decrease in creatine and phosphatidylcholine levels,[99] which—as we saw in Chapter 3—both require SAMe for their synthesis.

Another common feature of ALS is a significant increase in homocysteine levels, which tend to rise as the disease progresses and symptoms worsen.[100] If this elevated homocysteine is not properly cleared through the transsulfuration pathway (see Fig. 3b) or the remethylation pathway (see Fig. 3a), it becomes oxidized, leading to the over-

---

[97] https://doi.org/10.1007/s12017-009-8089-7
[98] https://doi.org/10.1016/j.bbadis.2013.10.004
[99] https://doi.org/10.3390/biomedicines9121944
[100] https://pubmed.ncbi.nlm.nih.gov/22781953/

activation of glutamate receptors (NMDA) and subsequent neuronal death by excitotoxicity, as has been observed in ALS.[101]

Among the potential metabolic routes for homocysteine, the transsulfuration pathway (see Fig. 3b)—which produces the antioxidants glutathione and taurine—tends to be more hyperactive in ALS. This appears to be a defensive response to the high oxidative stress levels that are characteristic of the cells in ALS patients.[102] For this reason, ensuring proper methylation via SAMe and maintaining adequate levels of the nutrients needed for this pathway (which we'll cover in the practical section) should be considered a complementary strategy in any pharmacological treatment.

Multiple sclerosis

Multiple sclerosis (MS) is an autoimmune disease that affects the central nervous system (CNS). In this condition, the immune system mistakenly attacks myelin, the protective coating around nerves, causing inflammation and damage. This disrupts the transmission of nerve signals between the brain and the rest of the body, impairing neural communication. MS symptoms can vary widely and include fatigue, mobility issues, visual disturbances, and cognitive problems, among others.

In MS, significantly reduced levels of SAMe have also been observed, particularly during the relapsing-remitting phases, which results in a diminished capacity for methylation.[103] Additionally, a decrease in the antioxidant glutathione[104] and an increase in homocysteine levels[105] have been detected, both of which can worsen symptoms and contribute to disease progression.

Huntington's disease

Huntington's disease is a hereditary neurodegenerative disorder that affects the central nervous system (CNS). It is caused by a genetic

---

[101] https://doi.org/10.1159/000517964
[102] https://www.sciencedirect.com/science/article/pii/S0969996120303004#s0115
[103] https://doi.org/10.1016/j.neuint.2017.10.011
[104] https://doi.org/10.1080/1354750x.2017.1334153
[105] https://doi.org/10.15190/d.2021.14

mutation on chromosome 4, which leads to the abnormal production of the protein huntingtin (HTT). This defect results in the progressive death of brain cells, especially in areas related to movement, cognition, and behavior. Symptoms of the disease include involuntary movements, cognitive decline, and emotional disturbances, typically appearing between the ages of 30 and 50.

As with other neurodegenerative diseases, Huntington's disease is marked by a low SAMe/SAH ratio—that is, reduced levels of SAMe and elevated levels of SAH.[106] This significant increase in SAH suppresses the activity of protein arginine N-methyltransferase 6 (PRMT6), which reduces the methylation of the gene that codes for the HTT protein. This loss of arginine methylation has a negative impact on neuronal function, as it plays a crucial role in regulating axonal transport in neurons by influencing HTT function.

Moreover, Huntington's disease is also associated with a marked decrease in cysteine levels—a precursor of both glutathione and taurine. This deficiency leads to increased oxidative stress, which significantly contributes to the neurodegeneration observed in the disease. For these reasons, boosting SAMe levels is considered a promising therapeutic approach, as concluded in a 2021 study.[107]

TYPE 2 DIABETES MELLITUS

This is a metabolic disease that affects how the body uses and processes nutrients and energy, specifically how it handles glucose. It is one of the most significant diseases in terms of both impact and prevalence. In 2020 alone, it was linked to over 100,000 deaths worldwide (a figure that continues to rise each year), and approximately 11% of the adult population—and 38% of those over the age of 65—in developed countries have been diagnosed with the disease, which translates to 1 in every 9 people. Furthermore, type 2 diabetes mellitus significantly increases the risk of the three most common and

---

[106] https://doi.org/10.1093/hmg/ddad125
[107] https://www.sciencedirect.com/science/article/pii/S2211124721002941#sec4

deadly diseases: cardiovascular disease, cancer, and neurodegenerative disorders.

While lifestyle habits such as poor diet and physical inactivity are widely recognized factors in the development of diabetes, the disease is also associated with low global DNA methylation, which increases the risk and worsens its progression. In a study[108] conducted on individuals with diabetes or impaired glucose tolerance, researchers observed that, after a one-year follow-up, patients whose condition worsened—despite receiving the same lifestyle intervention—had lower levels of methylation at the start of the study compared to those whose condition improved, regardless of other risk factors such as smoking or alcohol use, and even though they followed the same dietary pattern.

Subsequent studies have shown that SAMe may improve insulin resistance mediated by tumor necrosis factor alpha (TNF-α), a key marker of inflammation, suggesting it could play a beneficial role in treating type 2 diabetes mellitus.[109] In addition, SAMe has been shown to reduce levels of glucose, insulin, triglycerides, and oxidative stress —especially when combined with alpha-glycerylphosphorylcholine (α-GPC),[110] a precursor of choline (see Fig. 3a), which I discussed in my previous book, *Evolutionary Nutrition*. It has also been observed that SAMe increases mitochondrial DNA density in skeletal muscle— an aspect typically impaired in people with diabetes.[111]

Obesity and diabetes are often linked to diets high in fructose, which also generate elevated oxidative stress in visceral fat tissue. In this context, human studies have found that supplementation with SAMe (at a dose of 20 mg per kg of body weight) reduces oxidative damage (by increasing glutathione levels), lowers inflammation (by reducing TNF-α), and decreases fat accumulation.[112]

---

[108] https://doi.org/10.4161/15592294.2014.969617
[109] https://doi.org/10.3858/emm.2010.42.5.036
[110] https://doi.org/10.1080/09168451.2018.1559721
[111] https://doi.org/10.1093/jn/137.2.339
[112] https://doi.org/10.22543/7674.42.P163171

DEPRESSION

Depression is one of those conditions that many people believe they understand, but only those who have actually experienced it truly grasp its depth. It goes far beyond fleeting sadness; it is a sorrow that sinks to the core, draining energy and interest in everyday life and leaving behind a lingering sense of emptiness. This loss of vitality is often accompanied by feelings of worthlessness or guilt, difficulty concentrating, and, in more severe cases, recurring thoughts of death or suicide.

According to reports by the World Health Organization (WHO) in 2022 and 2023, depression is one of the leading causes of disability worldwide, affecting 5% of the global population—nearly 400 million people. Over the past ten years, the incidence of depression has increased by 18%, and each year more than 700,000 people die by suicide.

Pharmacological Treatment and Neurotransmitters

The approach taken by public health systems is primarily pharmacological. Most of the available antidepressant drugs aim to increase the availability of monoamine neurotransmitters in the brain—serotonin, norepinephrine, and dopamine—which are the ones most implicated in depression. These medications work in two ways: (1) by slowing the breakdown of these neurotransmitters through inhibition of the enzyme that degrades them (monoamine oxidase, or MAO), or (2) by inhibiting their reuptake, meaning they prevent the neurotransmitters from being reabsorbed by the receiving neurons after delivering their signal.

A crucial mechanism that contributes to the effectiveness of these medications—when they do work—is their ability to bind to the TrkB receptor of brain-derived neurotrophic factor (BDNF) and enhance its signaling and activation.[113] BDNF acts like a "fertilizer" that promotes neuronal plasticity, strengthening the connections between neurons and encouraging the formation of new neural networks. This effect is particularly relevant to the antidepressant response, as BDNF

---

[113] https://doi.org/10.1016/j.cell.2021.01.034

not only influences the hippocampus, which is essential for memory and learning, but also the amygdala, which regulates emotions, stress responses, and mood.

Circadian Rhythms and Neurotransmitters

It has also been shown that properly regulating circadian rhythms —by getting more exposure to natural light during the day and less to artificial light at night—promotes the synthesis of neurotransmitters like serotonin, dopamine, and norepinephrine. In the retina, there is a type of ganglion cell that responds to two kinds of light input and sends signals to the hypothalamus, where our master circadian clock is located. This clock synchronizes the body's biological processes according to the time of day. These two types of light are: long-wavelength light (the orange and red hues of sunrise and sunset) and short-wavelength light (commonly known as blue light, which is most intense during the day). Essentially, these cells tell the master clock, "it's sunrise now, it's daytime now, it's sunset now, it's nighttime," and based on that input, the clock sends signals to the rest of the body, such as: "it's daytime—raise cortisol and blood pressure to energize you," or "it's nighttime—lower cortisol and release melatonin to get the body ready for rest."

We can begin to grasp the importance of this system—a mechanism that has evolved over millions of years in all living beings, from single-celled organisms to the most complex, like us. Yet in industrialized societies, we're doing exactly the opposite of what this system requires, and as a result, we are throwing these rhythms out of sync. During the day, when we should be exposed to bright, natural light, we stay indoors under artificial lighting or behind windows that filter out UV-B rays, reducing light intensity and distorting its spectrum. At night, when light exposure should decrease, we keep our homes lit and continue using screens and other electronic devices. In other words, we're getting too little natural light during the day and too much artificial light at night—the exact opposite of what the body needs and what it has evolved to handle.

Until just a few years ago, it wasn't known that these retinal cells also send signals to other parts of the brain, such as the amygdala and the habenula—both of which are involved in depression and mood regulation.[114] Among its various functions, the habenula modulates the neurotransmitters most strongly linked to depression, sending signals that determine when to release or suppress serotonin, dopamine, and norepinephrine. This means that poor circadian regulation has a direct and negative effect on our mood and fosters depressive states. A 2023 study[115] showed that exposure to bright light at night increases the risk of depression, post-traumatic stress disorder, schizophrenia, anxiety, and bipolar disorder, while exposure to bright sunlight during the day reduces the risk of these conditions and improves mood and overall well-being.

Let's recap everything we've covered so far. Depressive states are marked by the following factors:

1. Decreased availability of monoamine neurotransmitters in the brain (serotonin, dopamine, and adrenaline): Increasing the levels of these neurotransmitters is the main target of the most commonly prescribed antidepressant medications, which work either by inhibiting the enzyme that breaks them down (MAO) or by preventing their reuptake by receptor neurons.
2. Decreased expression of BDNF, a type of brain "fertilizer" that promotes the growth, development, survival, and plasticity of neurons—and whose levels increase when antidepressant medications are effective.
3. Disruption of circadian rhythms, which in turn dysregulates the neurotransmitters mentioned above, worsening mood and increasing the risk of depression and other mental health disorders.

Now let's look at how these three factors relate to SAMe deficiency and, as a result, to reduced methylation capacity and epigenetic alterations.

---

[114] https://doi.org/10.1016/j.cell.2018.08.004
[115] https://doi.org/10.1038/s44220-023-00135-8#Sec3

SAMe Deficiency and Low Methylation

In 2014, the Department of Psychiatry at Taiwan University conducted a study that revealed a significant reduction in global DNA methylation (hypomethylation) in patients with depression compared to healthy individuals. Moreover, they found that the degree of methylation was inversely correlated with the severity of the depression—in other words, the lower the methylation, the more severe the depression, and vice versa.[116]

As we saw in Chapter 3, two of SAMe's critical roles as a methyl donor are the synthesis of monoamine neurotransmitters and the regulation of circadian rhythms. So, if we don't have enough SAMe available, our ability to properly synthesize neurotransmitters and regulate our circadian rhythms will be compromised—even if we try to alleviate symptoms with medication. Both processes are essential for maintaining a stable and healthy mood. When SAMe availability is low, the body's overall methylation capacity drops, and it will prioritize synthesizing only the minimal neurotransmitters needed for immediate survival—because, for the body, survival always takes precedence over functions like emotional well-being.

In 2022, the Departments of Molecular Biology and Neurobiology at the University of Israel first conducted a systematic review,[117] followed by a study,[118] in which they showed that SAMe increases monoamine neurotransmitter levels in the brain and reduces the activity of the enzyme MAO, which breaks them down. This alleviates depressive symptoms through epigenetic mechanisms—by modifying the expression of certain genes. They concluded that decreased global DNA methylation leads to aberrant expression of genes associated with depression, which ultimately drives the condition, and that increasing the availability of SAMe helps counteract this hypomethylation.

There are many more studies supporting these conclusions, but to avoid unnecessary length, let's briefly summarize the key findings from several of them:

---

[116] https://doi.org/10.2147/NDT.S71997
[117] https://doi.org/10.3389/fnbeh.2022.759052
[118] https://doi.org/10.3390/ijms231911898

- An excess of homocysteine—which reduces SAMe levels and causes hypomethylation—decreases serotonin synthesis and increases the activity of the MAO enzyme.[119]
- Low levels of vitamins D, B9, and B12—which, as we'll see in the practical section, raise homocysteine levels and promote hypomethylation—play a role in the pathogenesis of depression in children and adolescents.[120]
- Low expression and *silencing* of the BDNF gene due to hypermethylation results in the decreased BDNF levels observed in psychiatric and mood disorders.[121] As we've seen, global DNA hypomethylation caused by SAMe deficiency can lead to hypermethylation of specific genes like BDNF, contributing to mental illness. Current studies show that SAMe supplementation increases endogenous BDNF expression.[122]
- In individuals with depression who died by suicide, astrocyte dysfunction (astrocytes are central nervous system cells that support neuronal function) has been linked to abnormal methylation of certain genes. Cases with reduced astrocyte-associated gene expression (hypermethylation) also showed a pattern of global DNA hypomethylation.[123]
- Proper DNA methylation is a key epigenetic mechanism in major depressive disorder.[124]
- In cases of depression, increasing SAMe has been shown to be more effective than placebo, as effective as traditional antidepressant drugs, and more effective in combination with antidepressants than antidepressants alone.[125]

In addition to the direct links we've just reviewed, an excess of homocysteine—often the result of a disrupted cycle due to a deficiency

---

[119] https://doi.org/10.15761/JSIN.1000108
[120] https://doi.org/10.1111/camh.12387
[121] https://doi.org/10.1038/jhg.2013.65
[122] https://doi.org/10.1007/s12264-016-0023-z
[123] https://doi.org/10.1038/mp.2014.21
[124] https://doi.org/10.1007/978-3-319-53889-1_10
[125] https://doi.org/10.1186/s12991-020-00298-z

in one of the nutrients involved—not only reduces the body's methylation capacity by failing to regenerate SAMe, but also negatively impacts cardiovascular health.[126]

Why does this matter in depression? Because recent science has confirmed what classical thinkers had long suspected: the brain doesn't just influence the heart; the relationship is bidirectional. The heart also communicates with the brain via the vagus nerve, reaching the insular cortex, which plays a central role in emotional perception and regulation, influencing mood, behavior, and emotional responsiveness.[127] In other words, the heart is also an emotional organ—and a diseased heart can't transmit healthy emotions. Put simply: it's hard to have a healthy mind in an unhealthy body (*mens sana in corpore sano*). We are a *bodymind*, not a body and a mind functioning separately—they can be distinguished, but not understood in isolation.

So before turning to antidepressant drugs that block the enzymes breaking down serotonin or inhibit its reuptake—medications that come with a host of side effects—why not first consider using the natural molecules our own bodies produce? Molecules that might be in deficit, thereby limiting our ability to synthesize key neurotransmitters and BDNF, and that also help regulate our circadian rhythms?

AUTISM

More accurately referred to as autism spectrum disorder (ASD), this is a highly complex neurodevelopmental condition that affects communication, behavior, and social interaction. Symptoms and their severity vary widely from one individual to another—hence the term "spectrum." Still, there are two core characteristics: (1) challenges in communication and social interaction, such as difficulty understanding and using both verbal and nonverbal language, trouble forming relationships, and a limited ability to share interests and emotions; and (2) repetitive behaviors and restricted interests, such as rigid rou-

---

[126] https://doi.org/10.1186/1475-2891-14-6
[127] https://doi.org/10.1038/s41586-023-05748-8#Sec6

tines, intense focus on specific topics, repetitive actions, and unusual responses to sensory stimuli.

Although there is no single known cause and its origin is multifactorial—with a strong genetic component—recent evidence has shown that environmental factors, such as nutrient deficiencies and environmental pollutants, along with gene methylation and other epigenetic mechanisms (which are themselves influenced by environmental inputs), play a crucial role in the development of the disorder and in the quality of life of those affected. Let's explore this further.

Hyperhomocysteinemia

A common finding in children with autism, compared to those without the condition, is elevated levels of homocysteine, typically due to a deficiency in active forms of vitamins B9 and B12—both of which, as we've seen (see Fig. 3a), are essential for the breakdown of homocysteine and its conversion into SAMe. Furthermore, the higher the homocysteine levels, the more severe the symptoms of the disorder,[128] negatively affecting the development of adaptive behavior and communication skills in these individuals.[129] A 2022 study[130] involving young women with autism linked low levels of vitamin B9 with greater severity of the disorder and hypermethylation (silencing) of the BDNF gene—a gene, as we've already discussed, that plays a key role in brain development and becomes more active with SAMe supplementation.

These nutritional deficiencies, which lead to the buildup of homocysteine in the blood, result in two major consequences: (1) homocysteine is not converted back into SAMe (remethylation pathway, see Fig. 3a), leading to the characteristically low methylation capacity observed in autism; and (2) it is not converted into cysteine (transsulfuration pathway, see Fig. 3b), which is needed to synthesize the antioxidants glutathione and taurine—both of which are also found in very low levels in individuals with autism. Let's take a closer look at both aspects.

---

[128] https://doi.org/10.3389/fnmol.2022.947513
[129] https://doi.org/10.1016/j.psychres.2015.05.021
[130] https://doi.org/10.2217/epi-2021-0494

Global DNA Hypomethylation

Another hallmark observed in individuals with autism is globally reduced DNA methylation and increased oxidative damage when compared to control groups.[131,132] This decrease in methylation capacity leads to higher levels of inflammatory molecules and oxidative stress in the body—a condition that is further exacerbated by exposure to a well-known environmental toxin: phthalates found in plastics.[133] These compounds are not only endocrine disruptors but also inhibit methylation. We'll delve deeper into this in Chapter 8 (practical section), where we explore how to boost methylation capacity.

Antioxidants

Glutathione

Glutathione is a tripeptide composed of glutamate, cysteine, and glycine (see Fig. 3b). It exists in two forms: the reduced form (reduced glutathione [GSH]) and the oxidized form (oxidized glutathione [GSSG]). GSH is its active form and serves as the most important intracellular antioxidant. Through enzymes such as glutathione peroxidase (GPx), which requires selenium—making it essential to monitor blood selenium levels—GSH donates electrons to oxidative molecules like free radicals and peroxides, neutralizing them and protecting cells from oxidative damage. When GSH donates its electrons, it becomes oxidized, forming GSSG. The accumulation of GSSG is an indicator of oxidative stress within the cell, so the enzyme glutathione reductase (GSR) is responsible for recycling the oxidized form back into its reduced (active) form, GSH, to maintain its protective function.

We can imagine how critical it is for cellular health and proper biological functioning, especially as we age, to maintain adequate levels of reduced glutathione and a proper balance between its two forms. In individuals with autism, there is a pronounced imbalance between

---

[131] https://doi.org/10.1007/s10803-011-1260-7
[132] https://doi.org/10.3389/fped.2021.685310
[133] https://doi.org/10.3390/metabo13030458

GSH and GSSG—very low GSH and elevated GSSG—compared to healthy individuals.[134] This leads to significantly increased oxidative stress, neuroinflammation, mitochondrial dysfunction, and impaired regulation of glutamate receptors (an excitatory neurotransmitter), which is another hallmark of autism and is also influenced by a deficiency in the other key antioxidant, taurine.

Taurine

Taurine is an amino acid that can be obtained through diet as well as, like glutathione, synthesized endogenously from cysteine (see Fig. 3b), which in turn is recycled from homocysteine. Unlike the classic amino acids, taurine is not incorporated into proteins during their synthesis; instead, it primarily acts as a neurotransmitter and as a modulator of calcium and sodium channel activity in cells. It also functions as an antioxidant, but unlike glutathione, its action occurs outside the cell, where it helps protect against damage caused by ROS—those reactive molecules that contribute to oxidative stress and inflammation.

Although, as mentioned, the causes of autism are still not fully understood, in addition to what has already been discussed, studies have identified several disrupted pathways in different areas of the brain, with two standing out in particular: (1) the excitatory glutamate neurotransmitter system, and (2) neuronal inflammation.

1. **Glutamate neurotransmitter**: When glutamate signaling doesn't function properly, it can damage neurons through a process known as *excitotoxicity*, which occurs when too much calcium enters the neurons due to overactivation of their channels from excessive receptor stimulation. This triggers a cascade of problems: neuronal oxidative stress, increased permeability of the blood-brain barrier (which normally filters out substances that shouldn't reach the brain), and mitochondrial damage. Additionally, since glutamate plays a crucial role in the early devel-

---

[134] https://doi.org/10.1016/j.freeradbiomed.2020.07.017

opment of the cerebral cortex in children, its dysfunction can lead to abnormal development of this brain region.

2. **Neuroinflammation**: On the other hand, inflammation of the neurons occurs due to certain inflammatory processes, such as those mediated by NF-κB. This inflammation is triggered by substances like short-chain fatty acids, which can leak into the brain when there are issues like a leaky gut and increased blood-brain barrier permeability—two common characteristics observed in individuals with autism.

Taurine acts on both pathways. On one hand, taurine supplementation matches—and even outperforms—the effect of two standard drugs used in autism to counteract neuronal excitotoxicity caused by excessive glutamate signaling,[135] as it can bind to NMDA receptors and modulate the excessive influx of calcium into neurons. On the other hand, taurine substantially improves the neuroinflammatory process in autistic individuals. It also inhibits excessive microglial activation (which we'll address in the next section), induces autophagy, and promotes neurogenesis.[136]

Glycine

Let's now explore the role of glycine in this context of neuroinflammation in brains with ASD and excessive microglial activation. As we mentioned in the Alzheimer's section, microglia are the brain's macrophages (a type of white blood cell) that, in addition to cleaning up by removing damaged cells, debris, or redundant neurons and synapses, also attack invading pathogens through inflammatory responses. They're like patrolling officers who only "draw their weapons" (trigger inflammation) when necessary. However, when macrophages launch an inflammatory response without a real infection, they can damage healthy tissue.

---

[135] https://doi.org/10.1016/j.ibneur.2023.08.2191
[136] https://doi.org/10.3390/neuroglia4010001

This is where glycine comes in. Like taurine, glycine regulates macrophage activity, including that of brain microglia, preventing unnecessary activation and helping balance the innate immune response.[137] As I noted in my previous book *Evolutionary Nutrition* and discussed in a dedicated article[138] on my blog, there's a widespread glycine deficiency in the population. Increasing glycine intake helps prevent the unnecessary activation of microglia—a feature found in people with autism and in other conditions like Alzheimer's, Parkinson's, schizophrenia, and ALS, which we'll discuss later.

Glycine also protects intestinal epithelial cells against oxidative inflammatory stress, reduces intestinal inflammation, and exerts immunomodulatory effects, which may offer therapeutic potential for individuals with autism.[139]

Additionally, glycine is needed to synthesize glutathione.

Although it goes beyond the main focus of this book, we cannot overlook other nutritional compounds that may also influence this condition—for better or worse.

Let's take a look at some of them.

Vitamin D

Among its many functions, vitamin D supports neuronal proliferation, synaptic plasticity, neurotransmitter regulation, and the reduction of neuroinflammation and neurotoxicity. Studies have shown that children with autism tend to have lower levels of vitamin D compared to neurotypical children, and the lower the levels, the more severe the symptoms. Supplementation with vitamin D3 has been shown to significantly improve symptoms of the disorder,[140] especially when combined with omega-3s,[141] which enhance its anti-inflammatory effect.

---

[137] https://doi.org/10.1136/bmj.k1674
[138] https://www.curroclavero.com/glicina-un-aminoacido-imprescindible-en-el-que-todos-somos-deficitarios
[139] https://doi.org/10.3389/fendo.2019.00247
[140] https://doi.org/10.1080/1028415x.2015.1123847
[141] https://doi.org/10.1080/1028415X.2018.1557385

Gluten

Eliminating gluten from the diet may also be a worthwhile strategy to try. A 2023 systematic review[142] concluded that there is a modest comorbidity between autism and celiac disease, and that the existing body of evidence supports reasonable hypotheses suggesting gluten may have a generally adverse effect by worsening symptoms and quality of life in children with autism. In my view, there's nothing to lose by trying it—the risk is nearly zero (completely zero if gluten is removed) and the potential benefit could be significant. It also makes biological sense, as gluten increases the production of zonulin and exorphins (which we'll discuss next), both of which damage the gut by increasing its permeability and promoting inflammation—two common features of autism.

A1 Cow's Milk

The same applies to A1 cow's milk. There is evidence suggesting that it could exacerbate—or even trigger—symptoms in individuals with autism. I won't go into detail here, as I already covered this topic in my previous book, *Evolutionary Nutrition*, and in a blog article.[143] To summarize: A1 cow's milk—the type most commonly consumed today—is not the original form of cow's milk, which is more similar to human milk (A2 milk). Instead, it's the result of a genetic mutation and releases a bioactive molecule called beta-casomorphin-7 (BCM-7). Both BCM-7 and gliadorphin-7 (GMP-7), a peptide derived from gluten's gliadin, are opioid peptides often referred to as *exorphins* (short for exogenous morphine-like substances). These exorphins can cross the blood-brain barrier and bind to opioid receptors. Significantly higher concentrations of bovine BCM-7 have been found in children with autism compared to neurotypical children, and the higher the concentration, the more severe the symptoms.[144] Just like with gluten, switching from A1 cow's milk to A2 milk or to goat or sheep dairy products poses no risk or harm—so there's nothing to lose by trying it.

---

[142] https://doi.org/10.3390/nu13020572
[143] https://www.curroclavero.com/leche-de-vaca-a2-y-recomendaciones-de-lacteos
[144] https://doi.org/10.1016/j.peptides.2014.03.007

ADDICTIONS

We all have a general idea of what addiction is: a condition in which a person feels compulsively drawn to a substance or activity, despite its harmful consequences. What's less commonly known, however, is the biological foundation behind it. Addiction arises from changes in the brain that heighten the desire—or perceived need—to repeat the experience of pleasure (or relief) provided by the substance or activity.

Dopamine: the neurotransmitter

Imagine your brain has a "like" button that gets activated every time you engage in activities such as eating or socializing—both essential for survival. This button triggers the release of the neurotransmitter dopamine, which makes us feel good, generates pleasure, and motivates us to repeat those activities. It's the "reward system," an ancient motivational mechanism through which the brain rewards us by saying, "That was great—let's do it again." All of this works perfectly… in an ancestral environment.

Fortunately, today our basic needs are mostly met. However, we now have access to many things that didn't exist in the ancestral setting where this dopamine-driven reward system evolved. We tend to think of addiction only in terms of illegal substances like cocaine, or legal ones like nicotine and alcohol, but there are many other forms of addiction: to video games, social media, the endless novelty of news (which is why it's so hard to stop scrolling), ultra-processed foods, pornography, gambling, impulsive shopping, and more. The problem is that these activities release much more dopamine than usual, creating a much more intense sensation of pleasure. Over time, the brain learns to associate the activity with that pleasant feeling (reinforcement learning) and adapts to the elevated dopamine levels, which leads to an increasing need for the substance or behavior just to feel the same pleasure. This makes it very difficult for the person to resist the urge and stop on their own.

Epigenetics

Now that we understand how this works, let's look at how it ties into our main topic: methylation and epigenetics. As we saw when discussing depression, a state of global DNA hypomethylation can reduce the synthesis of certain neurotransmitters—including dopamine. Scientific evidence suggests that individuals with low global methylation—and consequently lower baseline levels of dopamine—may have a greater predisposition to addiction due to an epigenetic basis in the brain's DNA.[145]

People with naturally lower dopamine levels may seek ways to compensate for this deficiency and to experience feelings of pleasure or satisfaction. This could make them more vulnerable to addictive substances or behaviors that produce an artificial, short-term spike in dopamine levels.

Knowing this, correcting DNA hypomethylation could be a strategy worth considering. Back in 2010, a study[146] showed that treating mice with methionine to increase SAMe levels and methylation capacity reduced cocaine-associated reward memories by downregulating the brain's reward response. Conversely, inhibiting methylation capacity enhanced those memories, heightening both the rewarding effects and the psychomotor-stimulating properties of the drug. In 2012, a Chinese university demonstrated that global DNA hypomethylation contributes to cocaine's rewarding effect, and that reversing this hypomethylation through methionine supplementation (a SAMe precursor) significantly diminished the drug's reward response.[147] The Department of Pharmacology at the University of Estonia found similar results, but using SAMe itself rather than methionine to restore methylation capacity.[148] This research,[149] led by the neuroscience program at the University of Florida, concluded that "drug-seeking behaviors are in part attributable to a DNA hypomethylation-dependent process, and increasing the availability of methyl groups (SAMe) reduces behavioral sensitization to drug-seeking."

---

[145] https://doi.org/10.3389/fgene.2022.806685
[146] https://doi.org/10.1038/nn.2619
[147] https://doi.org/10.1371/journal.pone.0033435
[148] https://doi.org/10.1017/S1461145713000394
[149] https://doi.org/10.1523/JNEUROSCI.5227-14.2015#sec-26

AUTOIMMUNE DISEASES

Autoimmune diseases encompass nearly a hundred conditions and affect approximately 10% of the population—a figure that continues to rise. Among them are rheumatoid arthritis, lupus (SLE), multiple sclerosis (MS), type 1 diabetes, Hashimoto's hypothyroidism, Graves' hyperthyroidism, inflammatory bowel diseases (such as Crohn's disease and ulcerative colitis), psoriasis, celiac disease, pernicious anemia, and atopic dermatitis, among others. Despite being distinct illnesses, they all share a common trait: a misjudgment by the immune system, which mistakenly attacks the body's own tissues (with the involvement of lymphocytes), confusing healthy tissue for pathogens.

Environmental factors, through epigenetic mechanisms, likely play a more significant role than genetics in the development of these diseases. This is evident in studies of monozygotic twins (genetically identical) that show a high rate of discordance—that is, the proportion of cases in which one twin develops the disease while the other does not.[150] These environmental factors are known to reduce methylation. Multiple studies have shown that patients with autoimmune diseases exhibit global DNA hypomethylation,[151,152] as well as hypomethylation in both T[153] and B lymphocytes,[154] when compared to healthy controls. As we discussed in Chapter 3, SAMe and the process of methylation play a crucial role in normal immune responses, as well as in the maturation, proliferation, activation, and differentiation of lymphocytes. For instance, a diet deficient in methyl donors can lead to abnormal early development of B cells, potentially triggering allergic or autoimmune disease.[155]

Hyperproliferation of T and B lymphocytes is one of the hallmarks of most autoimmune diseases, including rheumatoid arthritis, lupus (SLE), multiple sclerosis (MS), and type 1 diabetes mellitus. SAMe

---

[150] https://doi.org/10.1016/j.jaut.2011.11.003
[151] https://doi.org/10.1016/j.imlet.2010.10.003
[152] https://doi.org/10.1002/art.25018
[153] https://doi.org/10.3389/fcell.2021.757318
[154] https://doi.org/10.1016/j.clim.2020.108622
[155] https://doi.org/10.1159/000337290

significantly reduces this proliferation by modulating the immune and inflammatory response and, according to a study[156] from Yale University, offers a simple and safe approach to autoimmune syndromes. In addition to its ability to modulate this response, SAMe may also inhibit autoimmune activity through its antioxidant properties, by increasing glutathione levels.[157] However, for the homocysteine produced as a byproduct of SAMe activity to be converted into glutathione, active vitamin B6 is required (see Fig. 3b). This vitamin has been found to be deficient in patients with autoimmune diseases,[158,159] and it plays a key role in immunomodulation by inhibiting T lymphocytes and reducing the inflammatory response.[160] This topic will be discussed in more detail in the practical section.

ALLERGIES

Another immune-related condition in which B and T lymphocytes play a role is allergy. Whereas in autoimmune diseases T cells attack the body's own cells, in allergies they overreact to the presence of a harmless particle or "antigen," mistakenly identifying it as a threat.

In a process called sensitization, T cells—specifically, type 2 helper T cells (Th2)—help activate B cells to produce specific antibodies known as immunoglobulin E (IgE), which bind to harmless allergens. When the person is exposed to the allergen again, these IgE antibodies trigger the release of chemical substances such as histamine, which cause allergy symptoms. Therefore, although B cells are the ones that produce the specific antibodies that mediate allergic reactions, it is actually the T cells that play a central role in directing and amplifying the immune response toward an allergic reaction.

All the accumulated evidence supports the hypothesis that allergies are driven more by environmental or *epigenetic* factors than by genet-

---

[156] https://doi.org/10.3109/08916934.2012.732133
[157] https://doi.org/10.1016/j.bbadis.2020.165895
[158] https://doi.org/10.1016/s0022-2143(96)90008-6
[159] https://doi.org/10.1016/j.mehy.2011.10.021
[160] https://doi.org/10.1155/2019/3121246

ics. For example, a study[161] that examined epigenetic differences between healthy individuals and allergy sufferers found that allergic patients had different proportions of T cell subtypes compared to healthy controls. This difference was attributed to variations in DNA methylation during the differentiation of helper T cells. This makes perfect sense, as we've seen that epigenetic regulation plays a key role in the differentiation of T lymphocytes—and that SAMe is essential for this process.

We also know that the enzyme DNMT is responsible for promoting DNA methylation using SAMe, and that Th2 cells play a predominant role in allergies. Well, it has been shown that DNMT acts as a rate-limiting factor in restraining Th2 cell-mediated allergic inflammation.[162] Moreover, SAMe supplementation has been found to reduce and improve allergic airway inflammation[163] by suppressing oxidative stress and reducing both inflammation and fibrosis. As a result, the Division of Allergy and Immunology at Seoul National University[164] has suggested that SAMe may represent a promising new therapeutic agent.

Additionally, SAMe is required as a methyl group donor for the enzyme histamine N-methyltransferase (HNMT) to function properly (see Fig. 2 and Chapter 3). HNMT is one of the two enzymes responsible for degrading histamine; it acts in the brain to prevent histamine excess (which we'll discuss shortly in relation to migraines) and also plays a key role in regulating the respiratory system's response to histamine.

MIGRAINES

Migraine is another neurological disorder that has risen significantly in recent decades and now affects nearly 15% of the population. It is characterized by episodes of intense headache, often debilitating, and

---

[161] https://doi.org/10.1371/journal.pgen.1004059
[162] https://doi.org/10.1073/pnas.1103803109#sec-2
[163] https://doi.org/10.1172/JCI34378
[164] https://doi.org/10.1038/emm.2016.35

can be accompanied by nausea, vomiting, and extreme sensitivity to light and sound.

Several key features are present in a migraine-prone brain: (1) increased neuronal excitability, which causes the brain to overreact to stimuli (such as stress, certain foods, hormonal changes, weather shifts, poor sleep, etc.) that shouldn't normally trigger such a response; (2) chronically low levels of certain neurotransmitters like serotonin or proteins like BDNF; and (3) dilation and inflammation of specific blood vessels in the brain, which leads to the stimulation of pain fibers and the onset of migraine symptoms. These symptoms can heighten the sensitivity of the central nervous system (CNS), further increasing the brain's excitability and making it more vulnerable to future attacks, thereby creating a vicious cycle that perpetuates migraine episodes.

Pharmacological treatment is mainly focused on relieving symptoms. However, the more relevant question is: what triggers these phenomena in the brain that ultimately lead to pain?

Misinterpretation by the Neuroimmune System

Among the various theories, one that stands out as particularly compelling is advocated by some neurologists and has been popularized in Spain by neurologist Arturo Goicoechea. This theory suggests that the problem lies in a misinterpretation by the neuroimmune system. The system mistakenly identifies a harmless situation, stimulus, or scenario (the migraine trigger) as a threat to the brain. Once this misjudgment is made, it initiates its defensive strategies—such as blood vessel dilation and inflammation—which then lead to migraine symptoms.

The brain is constantly receiving information from both the external environment and the body's internal state, which it evaluates through neural networks to make decisions. Most of the time, the brain gets it right. But sometimes, it doesn't—and it mistakes something harmless for a threat, as happens during migraine attacks. According to Dr. Goicoechea, migraine is something the brain "learns." Over time, the surrounding context sensitizes neural networks, making these interpretive errors more likely and more frequent.

In other words, the environment—including a drug-centered medical focus, cultural influences, media messages, family attitudes, and even neurologists themselves (when they overemphasize triggers and symptoms)—*feeds* migraine episodes. It reinforces the brain's belief that something important and harmful is happening when, in fact, it's simply overreacting to a harmless stimulus (like eating chocolate, being exposed to sunlight, southern winds, stress, menstruation, etc.).

Although the topic is complex, just understanding the biology of what's happening—knowing that during a migraine attack, nothing actually harmful is happening in the brain (even if the neuroimmune system mistakenly believes it is)—is already a crucial step toward *unlearning* the migraine. Fortunately, the brain is dynamic and constantly changing, which allows for new learning and functional rewiring.

SAMe, BDNF, and Serotonin

Regardless of how much the current "cultural" context contributes to reinforcing a migraine-prone brain, research has also shown that migraine brains, compared to control groups without migraines, have chronically lower levels of the BDNF protein[165] and the neurotransmitter serotonin.[166]

We've already discussed both in relation to depression, autism, and Alzheimer's disease. As we saw, SAMe is required to increase expression of the BDNF gene and to synthesize serotonin. So it's not surprising that people who suffer from migraines also tend to have lower SAMe availability, which leads to a reduced capacity for DNA methylation.[167] These conditions create a neurological context that makes migraine episodes more likely—but they may also contribute to the neuroimmune system's misjudgment mentioned earlier.

As early as the 1980s, SAMe was found to relieve pain in patients with migraines,[168] although the mechanisms at the time were poorly understood and were thought to be related to increased serotonin

---

[165] https://doi.org/10.1186/s10194-017-0725-2
[166] https://doi.org/10.5214/ans.0972.7531.12190210
[167] https://doi.org/10.52340/jecm.2022.08.14
[168] https://pubmed.ncbi.nlm.nih.gov/3514492/

levels. More recently, in 2023, these mechanisms have become clearer.[169] It has been shown that reducing homocysteine helps decrease the frequency of migraine attacks, and that increasing SAMe levels alleviates pain by compensating for the global DNA hypomethylation observed in these patients. SAMe also reduces inflammatory markers (such as TNF-α) and oxidative stress (by boosting glutathione), and helps regulate both BDNF and serotonin.

The fact that, in people with migraines, a SAMe deficiency may be responsible for low baseline levels of serotonin and BDNF—predisposing their brain to a migraine attack—is just the beginning. Following this deficit, a surge and elevated levels of both serotonin and BDNF are often observed, likely as part of a compensatory mechanism.[170] During a migraine episode, the sudden release of these two molecules may represent the brain's attempt to restore balance and reduce pain, but this process can also trigger neuronal overstimulation, contributing to migraine symptoms. Chronically low serotonin availability may form the biochemical basis for migraine etiology, while a sudden spike in its release may be among the key events that culminate in a migraine attack.[171] Low serotonin levels, for example, can lead to blood vessel dilation, while its abrupt release as compensation can intensify pain perception. This rapid release may also result in a subsequent drop in serotonin levels, perpetuating a cycle of fluctuations that contribute to heightened sensitivity and ongoing pain. A similar mechanism seems to occur with BDNF; its disproportionate release not only contributes to pain but also keeps patients sensitized to future episodes.[172]

Therefore, in my view, the first preventive approach should be to restore cerebral homeostasis by correcting the SAMe deficit and reducing homocysteine in order to rebalance serotonin and BDNF levels and prevent a brain predisposed to migraine attacks—a brain that, as

---

[169] https://www.doi.org/10.26502/fjppr.067
[170] https://doi.org/10.1177/25158163231190292
[171] https://doi.org/10.1111/j.1468-2982.2007.01476.x
[172] https://news.utdallas.edu/health-medicine/neuroscientist-making-headway-in-fighting-migraine/

Dr. Goicoechea explains, is further shaped by today's cultural environment. For example, a recent 2024 study[173] shows that molecules that enhance BDNF—such as SAMe—are effective regulators for managing migraines.

FIBROMYALGIA

Fibromyalgia is another chronic disorder characterized by widespread musculoskeletal pain, fatigue, sleep disturbances, memory problems, and, at times, mood issues. The exact cause of fibromyalgia remains unknown, and thus, there is no definitive cure. However, in recent years, increasing attention has been given—once again—to the inadequate interaction between genes and the environment, as well as to changes in DNA methylation that affect the central nervous system (CNS), improper neurotransmitter release, the bidirectional communication between the brain and the immune system, and altered responses to stress and inflammation.

In this regard, a pattern of hypomethylated DNA has been detected,[174] which appears to influence the stress response, DNA repair, autonomic nervous system activity, the clearance of free radicals, neuronal inflammation, and central pain sensitization.[175,176]

ATTENTION DEFICIT HYPERACTIVITY DISORDER (ADHD)

Attention deficit hyperactivity disorder (ADHD) is a neurological condition characterized by difficulties in maintaining attention, impulsivity, and, in many cases, hyperactivity. It affects both children and adults, influencing their behavior in social, academic, and professional settings. Stimulant medications are the first line of treatment and the most commonly prescribed. These work by increasing levels of certain neurotransmitters in the brain, such as dopamine and norepinephrine,

---

[173] https://doi.org/10.2174/0115701808274096231207042744
[174] https://doi.org/10.1177/1744806918819944
[175] https://doi.org/10.3390/jcm10214992
[176] https://doi.org/10.1097/j.pain.0000000000000932

which helps improve focus and reduce impulsivity and hyperactivity. Examples of these medications include methylphenidate (Ritalin® and Concerta®) and amphetamines (Adderall®), which—as expected—can have many undesirable side effects, such as insomnia, decreased appetite, mood swings, headaches, and increased heart rate and blood pressure, among others.

ADHD is another condition in which a significant decline in global DNA methylation has been detected.[177] It has been shown to improve with SAMe supplementation,[178] with minimal and temporary side effects up to a dose of 1,600 mg/day,[179] which did not interfere with daily functioning.

AGING

We've saved the mother of all diseases for last: aging. It is the single greatest risk factor for nearly every illness. I don't consider aging a disease in itself, as some experts do; I see it more as a physiological state, another stage of life. It's not a disease—it's a fundamental law of physics. The second law of thermodynamics states that "the amount of entropy in the universe tends to increase over time." Entropy is a measure of disorder (chaos) or randomness in a system, and if it inevitably increases over time and affects all systems, those systems undergo a gradual loss of order and structure, which leads to the accumulation of damage and, consequently, deterioration and dysfunction until they eventually collapse (death).

Faced with this reality, every system has two options: (1) to surrender and succumb to entropy, or (2) to draw energy from its environment to resist it, attempt to overcome it, and prevent its own destruction. The latter is the choice made by all living systems, which always strive to maintain themselves in the most orderly state possible. In humans, the body—as a complex system—fights to ensure self-preservation and self-replication. It constantly seeks to repair and protect

---

[177] https://doi.org/10.1038/s41380-022-01493-y
[178] https://pubmed.ncbi.nlm.nih.gov/2236465/
[179] https://doi.org/10.1007/s00702-012-0831-x

itself, becoming increasingly efficient at harnessing energy from its environment for this purpose. However, with our modern lifestyle (poor diet, caloric excess, physical inactivity, poor sleep hygiene, chronic stress, exposure to environmental toxins, etc.), this task is becoming increasingly difficult.

Epigenetics

As for the subject at hand, we've already discussed how epigenetic alterations and genomic instability are two of the aging-related causes identified by Carlos López-Otín. Both contribute to a more chaotic and disordered state in gene expression and cellular function, reflecting an increase in biological entropy. This growing molecular and genetic disarray leads to cellular dysfunction, compromising the integrity of tissues and organs and thus contributing to the overall aging process. Viewing aging through the lens of entropy offers a compelling framework for understanding it—it emphasizes the importance of preserving genetic and epigenetic information to maintain cellular function, slow down aging, and prevent related diseases. This book is precisely about exploring interventions that can help sustain—or even restore—healthy epigenetic patterns, equipping us with stronger tools to actively combat the *rise in entropy* and its devastating consequences.

SAMe Deficiency and Hypomethylation

Today, there is no doubt that global DNA methylation levels and aging are closely linked. As we've already mentioned, the general trend as we age appears to be the development of: 1) global DNA hypomethylation (in non-gene-controlling regions), which results in less efficient gene regulation, and 2) regions of hypermethylation in CpG islands (specific to gene promoters), leading to the inappropriate silencing of genes that should remain active. These epigenetic changes destabilize the genome and contribute both to aging[180,181] and to age-related diseases.[182]

---

[180] https://doi.org/10.1089/rej.2012.1324
[181] https://doi.org/10.1007/s11357-011-9216-6
[182] https://doi.org/10.1016/j.mad.2015.02.002

In addition, global DNA hypomethylation has been linked to a state of frailty, regardless of chronological age.[183] In other words, a younger person may be more physically fragile than an older one if they exhibit a more pronounced decline in global methylation. This is because they would have less epigenetic control, which is more closely tied to functional decline than to a person's chronological age.

On the other hand, studies have shown that a dietary intake that boosts methylation capacity—by increasing the availability of SAMe as a methyl group donor—can play a key role in maintaining DNA methylation. This could help protect the body from environmental factors that reduce methylation (such as radiation) and contribute to aging and overall deterioration.[184]

We're just about to reach the practical section, where we'll go over how to achieve this. But first, let's take a look at the few diseases in which the body presents the opposite of what we've seen so far: global hypermethylation due to excess SAMe.

KEY POINTS

1. Global DNA hypomethylation, caused by the loss of SAMe, is a common factor in a wide range of diseases. It compromises genomic stability, gene expression control, and cellular function, contributing to the development and progression of numerous conditions. Increasing SAMe levels to enhance methylation capacity may serve as an effective preventive and therapeutic strategy across these conditions.

2. **Cardiovascular disease**. The leading cause of death worldwide, with atherosclerosis as the core process, marked by the buildup of fatty plaques in the arteries. Factors like hyperhomocysteinemia and hypomethylation increase the risk of heart attack and stroke.

---

[183] https://doi.org/10.1007/s11357-011-9216-6
[184] https://doi.org/10.1016/j.cbi.2009.11.010

3. **Cancer**. Global hypomethylation promotes genetic instability and the silencing of tumor-suppressor genes, accelerating cancer progression and metastasis.

4. **Neurodegenerative diseases**. In Alzheimer's disease, hypomethylation facilitates the accumulation of beta-amyloid plaques and tau dysfunction. In Parkinson's, it reduces the effectiveness of treatment and promotes the overproduction of alpha-synuclein. In ALS and MS, it worsens neurodegeneration.

5. **Type 2 diabetes**. Hypomethylation is linked to disease progression, impacting insulin resistance and increasing oxidative stress and glucose levels.

6. **Depression**. Low methylation interferes with the synthesis of key neurotransmitters like serotonin, worsening depressive symptoms. Restoring methylation improves mood.

7. **Autism**. Elevated homocysteine and low methylation capacity worsen symptoms and impair neurological development. Improving methylation is key to enhancing well-being in affected individuals.

8. **Addiction**. People with global DNA hypomethylation may have lower baseline dopamine levels, making them more prone to addictive behaviors to compensate. Increasing SAMe helps regulate dopamine and reduces compulsive reward-seeking.

9. **Autoimmune diseases and allergies**. Hypomethylation of T and B lymphocytes amplifies immune responses in autoimmune conditions. In allergies, SAMe helps regulate inflammation and supports histamine breakdown, improving immune response.

10. **Migraines**. In migraine-prone brains, hypomethylation disrupts neurotransmitters like serotonin and BDNF, increasing suscepti-

bility to attacks. Restoring methylation can reduce the frequency of episodes.

11. **Fibromyalgia and ADHD**. Fibromyalgia is associated with hypomethylation, affecting pain response and inflammation. In ADHD, low methylation affects neurotransmitter regulation, and SAMe improves symptoms.

12. **Aging**. Global DNA hypomethylation increases with age, contributing to frailty and the decline in cellular function. Maintaining adequate methylation capacity is essential to slow biological aging.

CHAPTER 6
Diseases Related to Hypermethylation

Everything in the body seeks balance—homeostasis. Nothing is beneficial in deficiency, but neither is it in excess (as with most things in life), and the levels of S-adenosylmethionine (SAMe) are no exception. While we've discussed how SAMe levels tend to decline with age—reducing methylation capacity and contributing to aging and an increased risk of nearly every disease—an uncontrolled rise in SAMe (hypermethylation), although far less common, can also lead to serious health issues. That's why the body has a specific system—dependent on glycine—to eliminate excess SAMe, and it is critical that this system functions efficiently.

In this chapter, we'll examine the few diseases currently linked to hypermethylation that are supported by more substantial scientific evidence. However, this is an emerging field of research, and fortunately, it is gaining more attention. It is likely that in the future, hypermethylation will be connected to other diseases, particularly mental health disorders. As we'll explore in the practical section, it's essential to assess methylation status before implementing any interventions to increase SAMe. Boosting SAMe in someone already experiencing or predisposed to hypermethylation is dangerous—it could create a new problem or worsen an existing one.

SCHIZOPHRENIA AND BIPOLAR DISORDER

Although these are distinct conditions, I've grouped them together here because, in terms of epigenetics—and more specifically, methylation—they share more similarities than differences. The first thing we need to acknowledge is that both psychiatric disorders are so complex that we still don't fully understand them, as is the case with most brain-related conditions.

Schizophrenia is a psychotic disorder characterized by disruptions in thinking, perception, emotions, language, awareness of reality, and behavior. Symptoms are generally classified into three categories: "positive" symptoms (such as hallucinations and delusions), "negative" symptoms (including apathy, social withdrawal, and impoverished speech), and "cognitive" symptoms (such as problems with attention and memory).

Bipolar disorder, on the other hand, primarily affects mood, alternating between episodes of *mania* (euphoria, hyperactivity, grandiosity) and *depression* (sadness, disinterest, fatigue). The duration of these mood cycles can vary widely, ranging from rapid shifts to episodes that last for weeks or even months.

While it's true that both conditions have a strong genetic and hereditary component, it's equally true that, in approximately 50% of identical twins, only one develops the disease. This indicates that epigenetics also plays a crucial role. In other words, environmental factors and life circumstances influence how genes are expressed in these individuals, helping determine whether the disorder manifests or not.[185]

Hypermethylation in the brain

A key shared feature in both disorders is the increased presence of SAMe—*hypermethylation*—in the brain, particularly in the prefrontal cortex.[186] As a result, two critical genes, *GAD67* and *RELN*, are more silenced than they should be, meaning their expression is lower than optimal.[187] *GAD67* is essential for producing gamma-aminobutyric acid (GABA), a neurotransmitter that helps calm brain activity. Reduced *GAD67* expression may lead to lower GABA levels, potentially making the brain overly active or restless. Meanwhile, *RELN* plays a vital role in organizing the development and structure of brain cells. When *RELN* function is impaired, the brain's structural organization may

---

[185] https://doi.org/10.1016/j.psychres.2021.114218
[186] https://doi.org/10.1097/wnr.0b013e32800fefd7
[187] https://doi.org/10.1001/archpsyc.57.11.1061

suffer, negatively impacting overall brain function. In short, insufficient levels of these proteins may contribute to the symptoms of these mental disorders.

Global hypomethylation in the blood

Another characteristic is the global hypomethylation of DNA in the blood, compared to healthy individuals. Moreover, the earlier the disease manifests—a marker of severity—the lower the methylation levels. One of the beneficial effects of antipsychotic medication appears to be its ability to increase DNA methylation.[188] This hypomethylation in the blood is accompanied by low levels of betaine (TMG), which likely contributes to the overall hypomethylation by impairing the conversion of homocysteine into SAMe. This pattern has been observed in patients with schizophrenia and those with type I bipolar disorder (with a stronger manic component), but not in individuals with type II (with a more depressive tendency).[189]

Unlike other conditions where the methylation profile is more straightforward, the coexistence of hypermethylation and hypomethylation in different tissues (brain and blood) underscores the enormous complexity of these psychiatric disorders. In any case, even though global hypomethylation is present, increasing SAMe in these patients should be avoided.[190,191] In fact, the strategy should be exactly the opposite. In the practical section (see Chapter 9), we'll explore how to approach this.

Homocysteine

In addition to reducing SAMe levels, a second strategy for these patients should be lowering homocysteine levels. Elevated homocysteine is a well-documented risk factor in the scientific literature. Meta-analyses show that for every 5 µM increase in homocysteine concentration, the risk of developing schizophrenia rises by as much

---

[188] https://doi.org/10.1096/fj.11-202069
[189] https://doi.org/10.1016/j.pnpbp.2019.109855
[190] https://doi.org/10.1136/bcr-2018-224338
[191] https://doi.org/10.1093/ijnp/pyv054

as 70%,[192,193] especially in relation to positive symptoms.[194] Moreover, high homocysteine not only raises the risk of the disease—it has also been shown that specific supplementation to reduce homocysteine levels improves negative (manic) symptoms in individuals with schizophrenia after 16 weeks of treatment compared to a placebo.[195]

One likely reason why elevated homocysteine is problematic in these patients is its detrimental effect on white matter in the brain. White matter consists of nerve fibers insulated by a substance called myelin, which enables the fast and efficient transmission of electrical signals between neurons. When white matter is compromised, communication between different brain regions is impaired. In other words, the networks that allow various brain areas to interact are disconnected and fail to operate properly.[196] This disruption can affect information processing, coordination, and the integration of brain functions, contributing to the core symptoms of schizophrenia—disordered thinking, perception, and behavior.

For all these reasons, supplementation with the right vitamins to reduce homocysteine—as we'll discuss in Chapter 10—has been shown to significantly reduce psychiatric symptoms and shorten the duration of illness in people with schizophrenia.[197]

EPILEPSY

Epilepsy is a neurological disorder characterized by the brain's tendency to generate recurrent seizures. These seizures are sudden changes in the brain's electrical activity and can present in various forms—from brief episodes of staring or blanking out to intense jerking of the arms and legs.

---

[192] https://doi.org/10.1038/sj.mp.4001746
[193] https://doi.org/10.1186/s12881-015-0197-7
[194] https://doi.org/10.1016/j.clinbiochem.2019.12.003
[195] https://doi.org/10.1001/jamapsychiatry.2013.900
[196] https://doi.org/10.1038/s41537-024-00458-0
[197] https://doi.org/10.1017/S0033291717000022

Although the exact causes of epilepsy are still not fully understood, for a long time genetics was believed to be the primary factor. However, over the past decade, the focus has shifted toward epigenetic causes. Early studies in this area identified a different gene expression pattern in people with epilepsy compared to healthy individuals. Unlike in other neurological diseases, epilepsy has been associated with an increase in DNA methylation (hypermethylation),[198] which induces changes in neuronal excitability in epileptic brains.[199]

One of the mechanisms through which the ketogenic diet proves effective for patients with epilepsy is its ability to increase adenosine levels,[200] a compound that acts as a DNA methylation reducer (see Fig. 2). This effect helps prevent disease progression by decreasing both the frequency and severity of seizures.

Unfortunately, antiepileptic drugs, which are designed to suppress neuronal hyperexcitability, are ineffective in more than 30% of patients. Moreover, these medications do not address the comorbidities associated with epilepsy, nor do they prevent the disease's development and progression. Therefore, the most promising approach lies in the prevention of epilepsy and the slowing of its progression.

In recent years, numerous studies have convincingly shown that global DNA hypermethylation is involved in the development and maintenance of epilepsy. Biochemical strategies aimed at reducing this excess methylation—such as the ketogenic diet to boost adenosine levels and the use of glycine to absorb excess SAMe (methylation)—have proven more effective than current pharmacological treatments.[201]

This should come as no surprise, given that pharmaceutical interventions are a relatively recent development in the context of our complex evolutionary history. In contrast, *primordial metabolites* like adenosine and glycine, present since the earliest stages of life, are fundamental to all forms of biological existence.

---

[198] https://doi.org/10.1007/s00401-013-1168-8
[199] https://doi.org/10.1172/JCI65636
[200] https://doi.org/10.1016/j.neuropharm.2015.08.007
[201] https://doi.org/10.3389/fnmol.2016.00026

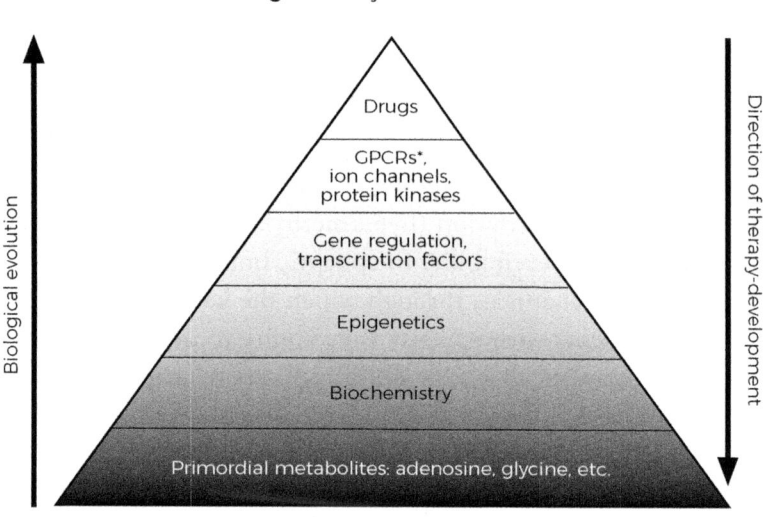

Figure 4. Pyramid of Life

Note. Evolutionary complexity started with key metabolites and biochemical mechanisms, which form the basis of all forms of life. In contrast, conventional drug development follows a top-down approach.

* GPCRs, G protein coupled receptors.

Source. Boison D (2016). The Biochemistry and Epigenetics of Epilepsy: Focus on Adenosine and Glycine. *Front. Mol. Neurosci.* 9:26. https://doi.org/10.3389/fnmol.2016.00026

ACUTE MYELOID LEUKEMIA (AML)

Acute myeloid leukemia (AML) is a type of cancer that affects the cells in the bone marrow—the soft tissue inside the bones where blood cells are produced. In AML, immature cells (myeloid progenitors), which would normally develop into red blood cells, white blood cells, or platelets, turn into abnormal leukemic cells known as blasts. These blasts do not function properly and can multiply rapidly, crowding out healthy blood cells and leading to symptoms such as fatigue, recurrent infections, bleeding, or bruising.

In relation to our topic, reducing methylation capacity—and therefore levels of SAMe—appears to be one of the few cases where this reduction is actually beneficial. As early as 2010, it was shown that patients with leukemia under the age of 65 who had lower global DNA methylation exhibited better clinical outcomes, including higher rates of complete remission, longer overall survival, and longer disease-free survival. In other words, lower global DNA methylation was associated with a higher rate of complete remission.[202]

In fact, decitabine and azacitidine, two of the most commonly used drugs in the treatment of AML—particularly in older patients or those ineligible for more intensive chemotherapy regimens—work precisely by reducing global methylation. Patients who respond to these drugs show a more pronounced decrease in methylation after treatment.[203] Furthermore, individuals treated with DNA methyltransferase (DNMT) inhibitors, which reduce methylation, experience significantly better overall survival and higher complete or partial remission rates compared to those receiving conventional care regimens.[204]

Given that epigenetic dysregulation lies at the core of AML, it stands to reason that reducing SAMe availability may be a promising complementary therapy alongside chemotherapy and other pharmacological treatments. This could improve the prognosis for patients with this disease. In this regard, a 2018 study[205] conducted by the Department of Hematology at the University of Cambridge found that reducing SAMe levels led to decreased growth and increased death of malignant leukemic cells. Most notably, in my view, this vulnerability was selective, affecting only cancerous cells while sparing healthy ones.

In 2019, researchers attempted to replicate these findings in mixed phenotype acute leukemia (with MLL-R), a particularly aggressive subtype of leukemia common in infants and children, which is associated with poorer overall prognosis and lower responsiveness to chemotherapy. The study[206] also showed that inhibiting SAMe significantly prolonged survival when combined with chemotherapy.

For all these reasons, nutritional strategies aimed at reducing SAMe and methylation—discussed in greater depth in the practical section—have proven highly beneficial. For example, eliminating the amino

---

[202] https://doi.org/10.1038/leu.2010.41
[203] https://doi.org/10.1182/blood-2012-05-429175
[204] https://doi.org/10.3390/ijms20184576
[205] https://doi.org/10.1182/blood-2018-99-115004
[206] https://doi.org/10.3390/cells8111322

acid methionine (the precursor to SAMe) from the diet slows disease progression.[207] Increasing vitamin B3 (niacin) not only reduces the severity of both short- and long-term chemotherapy side effects but also enhances the destruction of tumor cells.[208, 209] Likewise, metabolic ketosis is compatible with chemotherapy and increases DNA damage in leukemic blasts while providing protective effects for healthy lymphocytes.[210]

KEY POINTS

1. **Introduction to Hypermethylation**:
   - Maintaining balanced SAMe levels is essential. Both deficiency and excess can negatively impact health.
   - Hypermethylation is less common than hypomethylation but is associated with specific diseases.
   - Before intervening to increase SAMe, it's critical to assess an individual's methylation status.

2. **Schizophrenia and Bipolar Disorder**:
   - Common features: Both disorders exhibit hypermethylation in the brain and global hypomethylation in the blood.
   - Hypermethylation in the brain: Affects genes such as *GAD67* and *RELN*, leading to reduced GABA production and structural abnormalities in the brain.
   - Hypomethylation in the blood: This pattern correlates with the severity and age of onset of the disease. Antipsychotic medications have been shown to increase methylation levels.
   - Homocysteine levels: Reducing homocysteine is crucial, as elevated levels are linked to increased risk and severity of schizophrenia and bipolar disorder.

---

[207] https://doi.org/10.1182/blood.2022017575
[208] https://doi.org/10.1080/01635580701649628
[209] https://doi.org/10.1158/1535-7163.mct-09-0042
[210] https://doi.org/10.1097/01.HS9.0000974040.18942.3f

3. **Epilepsy**:
   - Characterization and causes: Epilepsy is associated with increased DNA methylation in the brain.
   - Dietary and epigenetic strategies: The ketogenic diet helps reduce hypermethylation by increasing adenosine levels, which in turn lowers the frequency and severity of seizures.
   - Limitations of antiepileptic drugs: Many patients are resistant to current medications. Epigenetic approaches such as ketogenic diets and glycine supplementation offer more promising preventive and adjunct therapies.

4. **Acute Myeloid Leukemia (AML)**:
   - Hypermethylation and prognosis: Lower global DNA methylation levels in AML patients are linked to better clinical outcomes and increased survival.
   - Therapies and drugs: Decitabine and azacitidine reduce global methylation and improve treatment responses and survival.
   - Strategies to reduce SAMe: Lowering SAMe and methylation levels proves beneficial, with interventions such as methionine-restricted diets and increased vitamin B3 intake enhancing chemotherapy effects and protecting healthy cells.

PRACTICE

(Theory Without Practice Is Useless)

CHAPTER 7
How to Know If You Have a State of Global Hypomethylation or Hypermethylation

I'm a deeply curious person, which makes learning—especially about foundational principles, the bedrock of all knowledge—something I truly enjoy. However, I consider myself even more practical and results-driven. Accumulating knowledge is meaningless if we don't apply it to solve real problems. That's why this is my favorite part: the practical. Let's dive in.

Is it possible to improve the state of methylation? Could we, at least partially, correct an imbalance that's either contributing to disease or accelerating the aging process?

Yes, it's possible—but before doing so, the first thing we need to know is our current methylation status, or *methylation index*.[211] In other words, we must determine whether our methylation capacity is within a normal range, and if it's not, whether we're dealing with a deficiency (hypomethylation), which would indicate a need to increase it, or an excess (hypermethylation), which would require reduction.

One might think that simply referring to Chapters 5 and 6—about the relationship between methylation status and specific diseases—would be enough to figure out whether we're hypomethylators or hypermethylators. But that's not sufficient. That's just the theory, and while it's useful, before applying any kind of intervention, it's essential to verify, case by case, whether an imbalance really exists, which direction it leans, and how significant it is. There are two ways to find this out, and both require a blood test.

---

[211] https://doi.org/10.1186/s13104-016-2296-8#Sec29

SAMe/SAH RATIO

We would need to determine the levels of two biomarkers: SAMe and S-adenosylhomocysteine (SAH) (see Fig. 2), and then calculate their ratio.

The ideal values are:

- SAMe: 95-120 µmol/l[212]
- SAH: < 12 µmol/l[213]
- Ratio SAMe/SAH: 8-12[214]

If your ratio falls within this range—even if it leans slightly toward one end or the other—there's no cause for concern. However, if the ratio drops below 8, it indicates low SAMe (and/or high SAH), which means there aren't enough methyl groups available to support proper methylation, leading to hypomethylation. On the other hand, if the ratio exceeds 12, that suggests an excess of SAMe, which favors hypermethylation.

This is the most reliable method because it directly measures how much SAMe your body has available for methylation, giving us a clear picture of your current state. The challenge, though, is that right now it's very difficult to measure these two biomarkers in most countries. Standard clinical labs don't offer this test due to lack of demand, as these markers are still largely unknown—even among healthcare professionals. In the U.S., while the test isn't widely available or inexpensive (around $230),[215] it *can* be done. But don't worry—if it seems too pricey, we have an alternative.

METHIONINE/HOMOCYSTEINE RATIO

The ideal values are:

---

[212] https://doi.org/10.1093/ajcn/nqab210
[213] https://doi.org/10.3389/fnut.2022.918698
[214] https://doi.org/10.1038/s41598-022-16242-y
[215] https://www.dhalab.com/shop/sam-esah-methylation-profile-plasma/

- Methionine: 27-58 µmol/l
- Homocysteine: < 8 µmol/l[216,217,218]
- Methionine/Homocysteine ratio: 3,4-8[219,220]

While this method is more indirect than the previous one, it's still quite reliable and commonly used in scientific studies. Methionine levels tend to correlate with those of SAMe, and homocysteine values align with those of SAH. Moreover, it's significantly more affordable —typically around $45-60—since both markers are more commonly measured in private labs.

If homocysteine is significantly elevated relative to methionine (ratio < 3.4), not only does it increase the risk of previously mentioned health issues, it also promotes the accumulation of SAH. This, in turn, reduces SAMe availability and downregulates the activity of the DNMT gene (the one responsible for DNA methylation), thereby encouraging hypomethylation. Additionally, elevated homocysteine levels suggest it's stuck in the cycle and not being recycled into SAMe, further diminishing SAMe availability and compounding the hypomethylation issue.

Conversely, if methionine is disproportionately high relative to homocysteine (ratio > 8), it likely indicates an excess of SAMe, since SAMe is synthesized from methionine—this favors hypermethylation.

**

Once we've determined whether an imbalance exists (and in my daily experience, I've found it to be far more common than most people think—especially when it comes to low methylation capacity), the next step is to correct it. The good news is that restoring balance doesn't require prescription medications. In the next two chapters,

---

[216] https://doi.org/10.1016/j.tjnut.2023.01.023
[217] https://doi.org/10.1038/s41598-017-05205-3
[218] https://doi.org/10.1155/2022/2156483
[219] https://doi.org/10.1001/jamapsychiatry.2019.1694
[220] https://doi.org/10.1016/j.clinbiochem.2019.12.003

we'll look at which supplements, foods, and other strategies can help address each type of imbalance—and why.

KEY POINTS

1. Although the previous chapters explain which diseases are associated with each type of methylation, it is essential to assess individual methylation status before implementing any strategy. You can't simply assume your methylation state based solely on the presence of certain diseases.

2. Theory must be complemented by blood tests for an accurate diagnosis. There are two main methods:
   - SAMe/SAH ratio: More direct, but also more expensive and less widely available.
   - Methionine/homocysteine ratio: Indirect but reliable, more affordable, and easier to test.

3. The lower the ratios, the lower the methylation capacity—and vice versa. As with everything in the body, the ideal is balance.

CHAPTER 8
How to Increase Methylation in a State of Global Hypomethylation

If a state of hypomethylation is detected—a fairly common condition, as we've seen—or if our levels fall within the normal range but we want to enhance our methylation capacity to counteract its natural decline with aging (which, in turn, increases the risk of nearly every type of disease), there are several strategies we can consider.

Let's take a look at them.

SUPPLEMENTAL SAMe

This is the most direct approach. S-adenosylmethionine (SAMe) is available as an over-the-counter supplement, but it's important to be selective when choosing the product. As is often the case with supplements, the cheaper ones tend to be low-cost for a reason: poor quality. Let's look at why.

SAMe supplements come in two forms: a biologically active form (S,S isomer), which is the same form produced and used by the human body, and an inactive form (R,S), which has no biological utility, doesn't participate in any metabolic processes, and therefore does not contribute to therapeutic effects. It's crucial to ensure that the supplement contains approximately 70-80% of the active (S,S) form. If this isn't clearly stated on the label, it's almost certain that the product contains the inactive form and wouldn't be worth the money. The active form is more expensive to produce, so if a brand invests in it, they'll usually highlight it on the packaging.

Additionally, the ideal supplement should have a pH-sensitive enteric coating—a protective layer that prevents it from breaking down in the acidic environment of the stomach. This ensures the supplement dissolves only when it reaches the more alkaline environment of the small intestine, which is where SAMe is absorbed.

As for the daily dosage, the typical recommendation ranges from 200 to 1,200 mg (and in certain severe or specific conditions, it can go as high as 3,600 mg), but I do not recommend exceeding 200 mg per day without professional supervision. Even though SAMe is sold over the counter, I view it as practically a medication, which means its supplementation must be approached with great caution. Excessive intake of SAMe (>8 mg of SAMe per kg of body weight) could be counterproductive and might even trigger the opposite of the intended effect—further DNA hypomethylation and disruption of circadian rhythms by lengthening their cycle.[221]

This happens because too much supplemental SAMe is broken down outside the methylation cycle into adenine—a known toxic metabolite and an inhibitor of adenosylhomocysteinase (AHCY) (see Fig. 2)—which promotes hypomethylation by causing SAH to accumulate.

To minimize these risks, it's best to ensure sufficient endogenous glycine availability (which, as we'll see in the next chapter, helps buffer excess SAMe) and to consider supplementing with it. For better absorption, SAMe should be taken on an empty stomach, and if more than one 200 mg capsule is consumed daily, it's preferable to split the dose into two separate intakes approximately 12 hours apart.

That said, unless there is a clearly pronounced global hypomethylation that can't be corrected with the compounds I'll describe next —or a very specific condition—I rarely use supplemental SAMe as a first option, and when I do, it's always at the lowest possible dose. I believe there are safer and more effective alternatives. Let's look at them.

CREATINE

One of my two favorite ways to boost SAMe levels and enhance methylation capacity is through creatine supplementation. Around

---

[221] https://doi.org/10.1038/s42003-022-03280-5#Sec16

40% of the SAMe available in the body is used for creatine synthesis. Therefore, by supplementing with creatine externally and saturating the body's stores, we relieve the system from the need to use SAMe for that purpose,[222] freeing it up for other vital functions.[223]

To calculate your daily creatine dose, multiply your body weight in pounds by 0.036 (0.08 if using kilograms); the result tells you how many grams you should take each day. For example, someone who weighs 150 lb should take about 5.4 grams per day. If the result exceeds 6 grams, it's advisable not to go beyond that amount.

Also, make sure that the creatine you use carries the *Creapure* label, which guarantees its purity.

Creatine has been widely studied and recognized as an ergogenic aid in sports, but its benefits extend far beyond physical performance. Personally, I find its effects outside the athletic context much more compelling. While there is growing discussion about its antioxidant and neuroprotective roles, its anticancer potential is still largely overlooked. Creatine may contribute to this effect through two main mechanisms:

1. By increasing the availability of SAMe, as we just discussed.
2. By supplying energy to CD8+ T cells with antitumor activity, which boosts anticancer immunity and enhances the effectiveness of other cancer immunotherapies through synergy.[224]

That said, before starting creatine supplementation—an aspect that's also rarely addressed—I prefer to ensure, through a blood test (as covered in a previous chapter), that the person does not have an issue with hypermethylation. Although creatine is one of the safest supplements available, precisely because it helps spare and increase SAMe availability, using it in individuals with excess SAMe (hypermethylators) could be highly counterproductive.

---

[222] https://doi.org/10.1371/journal.pone.0131563
[223] https://doi.org/10.3945/jn.111.144857
[224] https://doi.org/10.3390/nu13051633

PHOSPHATIDYLCHOLINE

My other favorite way to increase SAMe is by boosting phosphatidylcholine intake, based on reasoning similar to that of creatine. A very high percentage—also around 40%—of the SAMe available in the body is used to synthesize phosphatidylcholine. By taking it exogenously, we reduce the body's need to use SAMe for this function, preventing it from being "wasted." Additionally, there's a second way phosphatidylcholine helps increase SAMe availability: phosphatidylcholine is a precursor of choline, which in turn is a precursor of trimethylglycine (TMG), a compound that helps convert homocysteine into more SAMe (see Fig. 3a).

Because of these effects (increasing SAMe and choline), recent studies[225] have shown that phosphatidylcholine extends lifespan and reduces Alzheimer's risk by decreasing beta-amyloid toxicity.

Raising phosphatidylcholine levels is easy with soy lecithin, as it is very rich in this molecule. Taking between 6 and 12 grams per day (0.212 to 0.423 oz) provides about 1.5 to 3 grams (0.05 to 0.11 oz) of phosphatidylcholine. I recommend choosing soy lecithin that meets two key criteria: (1) Identity Preserved (IP), which ensures the product's traceability and purity through monitoring from cultivation to processing, and (2) Non-GMO (not genetically modified).

DIETARY CHOLINE

Another way to avoid wasting the SAMe available in the body is by increasing dietary choline intake. Phosphatidylcholine can be synthesized in two ways: one involves SAMe as a methyl donor—as already mentioned—and the other, known as the *Kennedy pathway* (see Fig. 3a), uses dietary choline. By increasing choline intake through food, we can support this second pathway, thus sparing SAMe from being used to synthesize phosphatidylcholine via the other route.

For example, one egg yolk contains about 150 mg of choline, so eating three yolks provides approximately 450 mg of choline per day,

---

[225] https://doi.org/10.1155/2019/2860642

which is the typical amount used in supplementation. Moreover, eggs in general—and choline in particular—offer numerous additional benefits that I covered in my previous book, *Evolutionary Nutrition*.

REDUCE OR AVOID ALCOHOL

Ethanol (ethyl alcohol) is the primary type of alcohol found in alcoholic beverages and is well known for being a toxin that affects the central nervous system. However, its epigenetic influence and its ability to alter methylation are less widely recognized. Alcohol reduces SAMe levels and increases SAH and homocysteine, thereby further lowering SAMe availability and methylation capacity.[226,227] In other words, through this pathway, it accelerates aging and increases the risk of virtually every type of disease.

Ethanol is a toxin under any circumstance, so the ideal scenario is to avoid it altogether. However, healthy individuals might consider reducing their intake to a *hormetic* dose—that is, an amount so small that the body can handle it, generating beneficial effects by activating its natural defense systems, without surpassing the threshold where adaptive responses are overwhelmed and systemic damage occurs. According to studies, this dose is limited to a maximum of one glass of red wine per day. I won't go into more detail here, as I also covered this topic extensively in my previous book, *Evolutionary Nutrition*.

DO NOT SMOKE

Another mechanism through which smoking causes cancer is by reducing the global DNA methylation capacity, leaving DNA unprotected and subsequently silencing specific tumor suppressor genes. One of the many compounds present in cigarettes is the carcinogen nicotine-derived nitrosamine ketone (NNK). It's important to note that this carcinogen is not found in the original tobacco plant but is formed during the curing and drying process of tobacco, as well as

---

[226] https://doi.org/10.1016/j.bbadis.2011.01.016
[227] https://doi.org/10.1093/jn/135.3.519

during its combustion when smoked. This carcinogen reduces the function of DNMT1 (see Fig. 2), the enzyme responsible for transferring the methyl group from SAMe to DNA, which decreases global methylation and leads to the hypermethylation (silencing) of tumor suppressor genes.[228,229] Remember that global DNA hypomethylation often leads to the hypermethylation of specific genes.

Interestingly, this effect is mitigated by our beloved, and sadly increasingly less consumed, Mediterranean diet. Among its many benefits, this diet has been shown to activate the AHRR gene,[230] a tumor suppressor gene that also protects against environmental toxins and helps detoxify foreign chemical compounds (xenobiotics).

EXERCISE

As expected, exercise is the best ally for healthy and functional aging. It's often said that exercise is the ultimate "polypill," benefiting virtually every system in the body—and I couldn't agree more. Beyond its well-known benefits, recent research has started exploring how exercise influences epigenetics.

In this field, long-term and consistent physical activity has been shown to increase global DNA methylation,[231] thereby reducing cancer risk.[232] In other words, exercise helps slow the age-related decline in methylation capacity and the associated overall deterioration. Both aerobic and resistance training (ideally both—there's no need to choose) have been proven to enhance methylation capacity and improve functional ability in older adults.[233] Additionally, exercise "silences" genes responsible for the development of Alzheimer's disease, acting as a protective factor for people over 50 and improving functional capacity, physical performance, and blood pressure.[234]

---

[228] https://doi.org/10.1172/jci40706
[229] https://doi.org/10.1038/s41598-018-23309-2
[230] https://doi.org/10.3390/ijerph20043635
[231] https://doi.org/10.1016/j.ejca.2013.02.013
[232] https://doi.org/10.1158/1055-9965.epi-18-0175
[233] https://doi.org/10.1113/EP089673
[234] https://doi.org/10.1016/j.exger.2024.112362

Another medium- and long-term benefit of exercise is its ability to increase the body's maximal oxygen uptake, or *VO2 max*—the maximum amount of oxygen the body can absorb, transport, and use within a given time. *VO2 max* is one of the most important indicators of physical fitness and a strong predictor of health and longevity. Research shows that as *VO2 max* increases, so does the expression and activity of the *RASSF1* gene, a major tumor suppressor that is inactivated in a wide range of cancers (breast, colon, endometrial, and others).[235]

I'll delve deeper into the relationship between exercise, methylation, and athletic performance in a special section of the appendix, since while exercise influences epigenetics, epigenetics also impacts physical performance.

SUFFICIENT AND QUALITY SLEEP

Another essential area that couldn't be left out is sleep, which until recently has been underestimated—though thankfully, its importance is increasingly being recognized. Sleep deprivation, or poor-quality sleep (meaning the failure to reach deep sleep stages or REM sleep), leads to general DNA hypomethylation,[236] which in turn promotes neurodegeneration and reduces neuroplasticity.

REDUCING ENDOCRINE DISRUPTORS: PHTHALATES AND BISPHENOL A

These compounds also promote DNA hypomethylation,[237] altering the epigenome and potentially affecting the offspring's characteristics when maternal exposure occurs during pregnancy.[238]

---

[235] https://doi.org/10.1371/journal.pone.0198641
[236] https://doi.org/10.1038/s41598-018-38009-0#Sec8
[237] https://doi.org/10.1186/s41021-022-00249-y
[238] https://doi.org/10.1073/pnas.0703739104

Phthalates

Phthalates are a group of chemical compounds used to make plastics—especially polyvinyl chloride (PVC)—more flexible and durable. They are found in a wide variety of products, including toys, food packaging, personal care products (cosmetics, perfumes, lotions, nail polish), medical supplies (IV tubing, blood bags), paints, varnishes, fabrics, and cleaning products. One study[239] shows that phthalates increase hypomethylation in autistic children and worsen the condition by also elevating oxidative inflammation.

Bisphenol A

Bisphenol A (BPA) is used in the production of plastics and resins and can be found in food and beverage containers, the inner coatings of cans and toys, and in thermal paper used for receipts (such as those from cash registers and credit card machines), among others. It is advisable to ensure that plastic products are labeled "BPA-free." Additionally, these items often have an identification code: a number inside a recycling triangle indicating the type of plastic. It's best to avoid those labeled with numbers 6 and 7.

**

Since, unless supplemented, all of the body's SAMe is produced from its precursor—the amino acid methionine (see Fig. 2)—some might wonder why increasing dietary methionine intake isn't included in the previous strategies for boosting endogenous SAMe synthesis. Logically, one might think that more methionine would enhance methylation capacity and compensate for losses due to genetics, aging, poor diet, or a combination of these. However, there are two main reasons why I'm not convinced by this strategy:

1. Methionine is found primarily in animal-based foods, especially meat. While it might seem logical that increasing methionine through higher meat consumption would raise SAMe

---

[239] https://doi.org/10.3390/metabo13030458

levels in the body, studies have shown that this is not the case. No significant differences in SAMe levels are observed when switching from a meat-rich diet to one that excludes meat.[240] This is because the body activates compensatory mechanisms to maintain balance when dietary methionine intake is reduced. SAMe production is not governed solely by methionine availability; it is also regulated by the activity of multiple enzymes that adjust production based on cellular needs. Methionine restriction actually improves metabolic efficiency, reducing waste and optimizing the use of available resources.

2. Furthermore, numerous studies have shown[241] that restricting methionine intake lowers the risk of age-related diseases and extends healthy lifespan. Among the many mechanisms behind this improvement are reduced fat accumulation in adipose tissue and the liver, enhanced insulin sensitivity and glucose utilization, optimized gut microbiota, reduced inflammation, increased antioxidant capacity that protects cells, enhanced autophagy, activation of tumor suppressor genes, and greater mitochondrial efficiency in oxygen use—along with an increase in mitochondrial number. Methionine restriction has also been linked to a reduced risk of dementia. I won't go into more detail here, as I discussed this topic more thoroughly in my previous book, *Evolutionary Nutrition*, along with the theory of "antagonistic pleiotropy" (a situation in which a mechanism or gene has beneficial effects for survival and reproduction early in life, but harmful effects in later stages).

Therefore, since restricting methionine in adulthood may actually be beneficial and, moreover, is not effective for increasing SAMe, I prefer not to rely on this strategy. Instead, I focus on enhancing SAMe through the other methods discussed in this chapter, as well as promoting the conversion of homocysteine into methionine and SAMe

---

[240] https://doi.org/10.3390/metabo12050373
[241] https://doi.org/10.1515/revneuro-2018-0073

—something we'll explore in Chapter 10—thus "killing two birds with one stone": eliminating the toxicity of homocysteine while simultaneously increasing SAMe availability.

KEY POINTS

1. **Strategies to boost methylation**:
   - Main goal: To counteract the decline in methylation that occurs with aging, which is linked to a higher risk of disease.
   - Importance of balance: Methylation should be increased cautiously, avoiding excesses that could lead to hypermethylation.

2. **SAMe** (supplementation):
   - Choosing the right supplement: Select products that specify at least 70-80% of the active S,S isomer and have enteric coating to ensure proper intestinal absorption.
   - Dosage and precautions: The recommended dosage is 200-1,200 mg per day on an empty stomach, but doses over 200 mg should not be taken without professional supervision. Excess SAMe can be counterproductive and may even promote hypomethylation.
   - Balance with glycine: Supplementing with glycine helps neutralize excess SAMe.

3. **Creatine**:
   - Preserving SAMe: Taking creatine supplements reduces the need for SAMe in its synthesis, thereby increasing its availability for DNA methylation and other non-epigenetic functions.
   - Dosage and purity: The recommended dose is 0.08 g/kg (0.036 g/lb) of body weight, without exceeding 6 g (0.21 oz) per day. Choose creatine with the *Creapure* label to ensure purity.
   - Additional benefits: Creatine offers neuroprotective, antioxidant, and potential anticancer effects by enhancing the function of antitumor CD8 T cells.

4. **Phosphatidylcholine**:
   * Boosts SAMe and choline levels: Intake of phosphatidylcholine reduces SAMe consumption during synthesis and increases choline, which is a precursor to TMG and helps convert homocysteine into SAMe.
   * Dosage and sources: Soy lecithin is a good source of phosphatidylcholine (6-12 g/day or 0.21-0.42 oz/day). Choose lecithin that is Identity Preserved (IP) and Non-GMO.

5. **Dietary choline**:
   * Stimulating the Kennedy pathway: Increasing choline intake through diet (e.g., egg yolks) supports phosphatidylcholine synthesis without using SAMe.

6. **Avoid alcohol**:
   * Negative epigenetic effects: Alcohol reduces SAMe and increases SAH and homocysteine, which lowers methylation capacity.
   * Safe intake: Limit consumption to hormetic doses (no more than one glass of red wine per day).

7. **Don't smoke**:
   * Epigenetic disruption by NNK: Smoking reduces global methylation and causes hypermethylation (silencing) of tumor-suppressor genes.
   * Protection via Mediterranean diet: This diet can mitigate the harmful effects of smoking by activating the AHRR gene.

8. **Exercise**:
   * Boosts global methylation: Regular physical activity enhances methylation capacity and reduces cancer risk.
   * Protective effects: Both aerobic and strength training improve functional capacity and lower the risk of Alzheimer's disease.

9. **Get enough quality sleep**:
   * Impact on methylation: Lack of sleep or poor-quality sleep reduces global DNA methylation, promoting neurodegeneration.

10. **Limit endocrine disruptors: phthalates and bisphenol A**:
    - Adverse epigenetic effects: Both phthalates and BPA promote DNA hypomethylation and may affect future generations' health.

11. **Do not increase dietary methionine**:
    - Endogenous SAMe regulation: Increasing methionine intake does not raise SAMe levels, as the body regulates production efficiently.
    - Methionine restriction: Limiting methionine improves metabolic health, reduces the risk of age-related diseases, and supports longevity.

CHAPTER 9
How to Reduce Methylation in a State of Global Hypermethylation

Since we gradually lose S-adenosylmethionine (SAMe) with age, it is far less common to find oneself in a state of global hypermethylation. However, it can occur due to genetic predispositions or nutritional imbalances and, as we saw in Chapter 6, an excess of SAMe is also harmful and has been linked to certain diseases. As mentioned in Chapter 7, before taking any action, we must first make sure that an imbalance exists and determine what kind. If we find there is a state of hypermethylation due to an excess of SAMe, fortunately there are several strategies available to minimize this imbalance and help restore homeostasis.

GLYCINE

Glycine is the first, most direct, primary, and essential strategy. The body's natural mechanism for "absorbing" and eliminating excess SAMe is through the action of the enzyme glycine N-methyltransferase (GNMT) (see Fig. 2). This enzyme transfers the excess methyl group from SAMe (the donor) to glycine (the acceptor), converting it into the amino acid sarcosine and transforming SAMe into S-adenosylhomocysteine (SAH). In this way, by consuming glycine and turning it into another molecule, the body disposes of surplus SAMe.

However, this process cannot occur unless we have enough glycine available in the body. As I mentioned in my previous book, *Evolutionary Nutrition*, and in a dedicated article[242] on my blog, we are all deficient

---

[242] https://www.curroclavero.com/glicina-un-aminoacido-imprescindible-en-el-que-todos-somos-deficitarios

in glycine—by about 10 grams (0.35 oz) per day,[243] even when we are in a state of normal global methylation, with no additional demand for glycine. This is because the functions of glycine in the body are so numerous and so essential that there simply isn't enough to go around: collagen synthesis, creatine synthesis, antioxidant glutathione production, heme group formation in hemoglobin, regulation of the nervous system (maintaining the balance between excitation and inhibition), ammonia detoxification, body temperature regulation during sleep, hormonal stimulation, glucose spike reduction, protection against osteoarthritis and joint degeneration, and more.

In addition to the body's high demand for glycine, modern dietary patterns have shifted away from those of our ancestors and the foods we evolved with over millions of years. The modern diet has eliminated many of the foods richest in glycine, such as animal skin, connective tissue (gelatin), bones (marrow and bone broth), and slow-cooked gelatinous cuts of meat (like shank, oxtail, beef or pork trotters, etc.).

In a state of global hypermethylation, enhancing the glycine pathway to eliminate excess SAMe—and prevent the damage it causes—has an added benefit. As we've seen, when glycine accepts a methyl group from SAMe, SAMe is no longer SAMe, and glycine becomes sarcosine, which indirectly increases levels of this amino acid. This is important because sarcosine inhibits glycine transporter 1 (GlyT1),[244] which removes glycine from the synaptic space in the brain—a transporter that is overexpressed in hypermethylation-related conditions such as epilepsy.[245] In other words, this transporter—normally kept in check by sarcosine—when overexpressed, excessively depletes glycine availability in the brain. Therefore, by increasing glycine exogenously (through supplementation and diet), we not only raise its levels directly but also indirectly, by reducing the activity of the transporter that removes glycine from the brain—something critical for these patients.

---

[243] https://doi.org/10.1007/s12038-009-0100-9
[244] https://doi.org/10.3389/fnmol.2020.00097
[245] https://aesnet.org/abstractslisting/glycine-transporter-1-mediated-suppression-on-epileptic-seizures--cell-type-contributions-and-mechanism

It's simply not possible to reach the blood glycine levels needed to counter a state of hypermethylation (and cover all its other vital functions) through food alone—so supplementation is essential. Those of you who've been following my work for a while know glycine is one of my favorite supplements. In powder form, it's inexpensive and has a naturally sweet taste, making it a great alternative to less healthy sweeteners like sugar. However, when dealing with something as serious as excess SAMe, it's better to control the daily dosage.

As for dosage, it's impossible to give a one-size-fits-all recommendation because it depends on several factors: the individual's body weight, their condition, the severity of the global hypermethylation, and their current blood glycine levels. That's why, especially in cases of serious illness, it's essential to get a blood test before starting supplementation.

One last note on glycine: although people with global hypermethylation require higher glycine levels due to increased demand, maintaining adequate glycine levels is important regardless of methylation status—whether hyper or hypo. There are two main reasons for this: (1) the wide range of crucial roles glycine plays, as we've already discussed, and (2) even in the more common state of SAMe deficiency (hypomethylation), if we use strategies to increase SAMe levels, we must ensure there's enough glycine available to absorb any excess and maintain homeostasis at all times.

VITAMIN B3

When hypermethylation is particularly elevated, another additional strategy is to use high doses of vitamin B3. The reason is that, when there's an excess of vitamin B3 in the body, in order to regulate its concentration and prevent toxicity, the body disposes of the surplus by methylating the vitamin—consuming SAMe in the process (via the enzyme nicotinamide N-methyltransferase [NNMT]) (see Fig. 2)—and excreting it through the urine. In this way, supplementing with a supraphysiological dose of vitamin B3 causes the body to consume a significant amount of SAMe to eliminate the excess, which can help reduce hepatic SAMe levels and the state of hypermethylation. This, in turn, may alleviate symptoms of conditions associated with excess SAMe, as has been demonstrated in cases of schizophrenia.[246]

Another added benefit of supplementing with vitamin B3 in these patients is that this vitamin increases levels of nicotinamide adenine dinucleotide (NAD+) because it serves as its precursor. NAD+ is a vital coenzyme in cellular metabolism, responsible for transporting electrons in processes such as cellular respiration and energy production in the mitochondria (adenosine triphosphate [ATP]). It also plays a crucial role in DNA repair, cell signaling, and the regulation of the circadian cycle. NAD+ levels decline with age, a process linked to various aging-related conditions and diseases. In disorders associated with excess SAMe, such as schizophrenia, NAD+ levels have been found to be significantly lower than in healthy controls.[247,248] In these patients, this NAD+ deficit amplifies an "immuno-oxidative" pathway, which includes oxidative stress, mitochondrial dysfunction, neurotoxicity, neuroinflammation, synaptic dysfunction, cognitive decline, and neuronal death. Supplementation with vitamin B3 has been shown to reverse these effects by raising NAD+ levels[249] and supporting clinical treatment in these individuals.

The two most common forms of vitamin B3 are niacin (or nicotinic acid) and nicotinamide (or niacinamide). For patients dealing with a state of hypermethylation, I recommend using nicotinamide, as this is the form in which excess vitamin B3 is methylated and excreted. Niacin, by contrast, requires intermediate steps to convert into nicotinamide before it can be methylated and eliminated, making it less efficient at reducing a global hypermethylation state. Additionally, high doses of niacin (>1,000 mg) are more likely to cause liver damage and toxicity.[250]

The typical dosage of nicotinamide used in these cases ranges from 500 to 1,000 mg per day, and it's best not to exceed that amount. If

---

[246] https://www.europeanreview.org/wp/wp-content/uploads/988-997.pdf
[247] https://doi.org/10.1093/schbul/sbw129
[248] https://doi.org/10.1038/s41398-023-02568-2#Sec22
[249] https://doi.org/10.1038/s41537-023-00357-w
[250] https://pmc.ncbi.nlm.nih.gov/articles/PMC1003029/pdf/westjmed00098-0084.pdf

the higher end of that range (1,000 mg) is needed in a specific case, the ideal approach is to split it into two separate 500 mg doses taken 12 hours apart.

KETOGENIC DIET

A third strategy involves following a ketogenic diet—that is, a diet very low in carbohydrates and high in fats, which induces ketosis, a metabolic state in which the body uses fat instead of carbohydrates as its primary energy source, producing ketones in the liver. Although in my previous book, *Evolutionary Nutrition*, I explained why I do not recommend a continuous ketogenic diet for healthy individuals, there are certain conditions in which it can be very beneficial—among them, those associated with global hypermethylation and excess SAMe, such as epilepsy.

What makes the ketogenic diet especially relevant in these cases is that it increases levels of β-hydroxybutyrate, one of the three ketone bodies produced by the liver. β-hydroxybutyrate is a specific type of ketone that, among other functions, raises adenosine levels in the hippocampus, which in turn increases SAH and blocks DNA methyltransferase (DNMT)—the enzyme responsible for methylating DNA (see Fig. 2)—thereby reducing global DNA methylation.[251,252] This mechanism helps explain why the ketogenic diet is so effective in epilepsy cases after the fourth week, as it reduces seizure activity by lowering DNA methylation.[253] The ketogenic diet is also being explored for other hypermethylation-related conditions such as schizophrenia and bipolar disorder, with very promising results according to a recent 2024 study conducted by Stanford University and other institutions.[254]

[251] https://doi.org/10.3390/nu14153245
[252] https://doi.org/10.1093/med/9780197501207.003.0025
[253] https://doi.org/10.1111/epi.17351
[254] https://www.sciencedirect.com/science/article/pii/S0165178124001513#sec0013

CHLOROGENIC ACID

Like the ketogenic diet, chlorogenic acid—a polyphenol found in many plants, especially coffee—has been shown to reduce global DNA methylation by increasing SAH levels and inhibiting the DNMT enzyme.[255]

Because of this hypomethylating effect, chlorogenic acid has shown benefits in hematologic neoplasms, exerting antitumor activity on T cells in cases of acute leukemia.[256]

To effectively increase chlorogenic acid intake for the purpose of reducing methylation, one should not only consume roasted coffee regularly and in moderate amounts (fewer than 5 cups a day), but also take green coffee extract supplements in doses ranging from 800 to 1,200 mg per day, divided into 2-3 doses of 400 mg each. It is important to ensure that the supplement contains at least 50% chlorogenic acid. This compound is also found in artichokes, pears, apples, blueberries, and in the skin of eggplants and potatoes.

One clarification: this does not mean that people who are looking to increase their methylation capacity should avoid drinking coffee. Moderate coffee consumption has a positive effect on health due to the complex interaction of its various components and their systemic actions, as I explained in detail in my book *Evolutionary Nutrition*. Also, the amount of chlorogenic acid present in moderate coffee intake is not sufficient to reduce methylation, which is why, as mentioned earlier, supplementation is necessary to achieve this effect.

KEY POINTS

1. **Strategies to reduce methylation**:
   - Main objective: To counter excess SAMe in cases of global hypermethylation in order to restore balance and prevent related diseases.

---

[255] https://doi.org/10.1093/carcin/bgi206
[256] https://doi.org/10.1590/1678-4685-GMB-2019-0347

- Importance of proper diagnosis: Before taking action, it's essential to determine whether a state of hypermethylation exists and identify its cause, in order to apply the appropriate intervention.

2. **Glycine**:
    - Eliminating excess SAMe: Glycine plays a key role in this process via the GNMT enzyme, which converts glycine into sarcosine.
    - Widespread glycine deficiency: Due to the lack of glycine-rich foods in the modern diet and the amino acid's many functions in the body, supplementation is recommended.
    - Supplementation and dosage: Doses should be individualized based on clinical needs and lab results, as glycine plays a critical role regardless of methylation status.

3. **Vitamin B3** (nicotinamide):
    - SAMe reduction: When taken in high doses, vitamin B3 is methylated by the body to eliminate the excess. This process uses up part of the excess SAMe, helping to lower its levels.
    - Additional benefit: Vitamin B3 increases NAD+ levels, helping to reverse damage associated with NAD+ deficiency in diseases linked to hypermethylation.
    - Recommended dosage: Between 500 and 1,000 mg of nicotinamide per day, divided into two doses for the higher end. Do not exceed this amount to avoid toxicity.

4. **Ketogenic diet**:
    - Effect on methylation: The ketogenic diet increases levels of β-hydroxybutyrate, a ketone that raises adenosine levels in the brain, thereby reducing DNMT activity and decreasing global methylation.
    - Targeted conditions: The ketogenic diet is especially helpful in cases such as epilepsy and is showing promising results in other hypermethylation-related diseases like schizophrenia and bipolar disorder.

5. **Chlorogenic acid**:
    * Reduction of global methylation: This polyphenol, found in coffee and certain plants, increases SAH levels and inhibits DNMT, thereby reducing methylation.
    * Intake and supplementation: To achieve significant effects on methylation, it is recommended to use green coffee extract supplements (800-1,200 mg/day), in addition to consuming roasted coffee and chlorogenic acid-rich foods.
    * Importance of balance: Moderate coffee consumption does not affect methylation and provides overall health benefits.

CHAPTER 10
How to Lower Homocysteine Levels

Regardless of whether we're dealing with the more common state of hypomethylation or one of hypermethylation, high levels of homocysteine (hyperhomocysteinemia) must always be addressed. Nothing —let me repeat, *nothing*—in the body benefits from elevated homocysteine levels. Quite the opposite, as we've seen throughout this book.

High homocysteine is not exclusive to hypomethylation; it can occur in both scenarios. Let's break each one down:

1. **Global hypomethylation**: When there's an imbalance favoring hypomethylation (low methionine/S-adenosylmethionine [SAMe] and high homocysteine), elevated homocysteine indicates a blockage in the process. In other words, homocysteine builds up in the blood instead of being recycled into methionine and then into SAMe. In this scenario, hyperhomocysteinemia would be the cause—or one of the causes—of the low methylation capacity, negatively impacting methionine/SAMe synthesis. The first step should be to lower homocysteine levels by evaluating which of the following supplements to use. After that, we can work on reversing the low methylation with the strategies and supplements discussed in Chapter 8.

2. **Hypermethylation**: Although less common, elevated homocysteine can also occur without an imbalance in the methionine/homocysteine ratio (i.e., both methionine/SAMe and homocysteine are high). In this case, the elevated homocysteine is a consequence of hypermethylation. Since homocysteine is an unavoidable byproduct of any methylation process, it can accumulate due to high methylation activity, especially if something

in its cycle is preventing efficient elimination. This pattern may be seen in diseases like epilepsy and acute myeloid leukemia, or even in conditions such as schizophrenia, bipolar disorder, and others—where hypomethylation is observed in peripheral blood (or leukocytes) while hypermethylation and high methionine coexist in other tissues like the brain,[257] along with elevated homocysteine levels.[258, 259] In these cases, the first step, just like with hypomethylation, is to address the hyperhomocysteinemia. But unlike that scenario, using methylation-promoting supplements like SAMe, creatine, and so on would be counterproductive.[260] Instead, it would be more appropriate to use methylation-reducing supplements such as glycine and vitamin B3, among others,[261, 262] as outlined in the previous chapter.

As we've already discussed, the body has three possible pathways (see Figs. 3a and 3b) for disposing of homocysteine and preventing its accumulation and the damage it can cause. In public health systems, this is a biomarker that is rarely requested—perhaps due to lack of awareness or because there are no specific pharmaceuticals to reduce its levels. And on the rare occasion it *is* tested, anything below 15 µmol/L is often considered normal. My approach here is much more aggressive: I don't like to see it go above 8 µmol/L. Higher levels not only offer no benefit, but as we've already thoroughly explained, they lead only to negative consequences. What's more, when the molecule gets stuck in the system, it can't be converted into other more beneficial substances like SAMe or antioxidants.

If blood levels exceed 8 µmol/L, it means that—whether due to genetic factors, poor diet, or both—something is off in one part of the cycle. This suggests that a process isn't functioning optimally and we need to identify and correct the root cause. The good news is that once

---

[257] https://doi.org/10.1007/s00702-004-0128-9
[258] https://doi.org/10.1016/j.clinbiochem.2019.12.003
[259] https://doi.org/10.1016/j.pnpbp.2019.109855
[260] https://doi.org/10.1093/ijnp/pyv054
[261] https://pubmed.ncbi.nlm.nih.gov/25855923/
[262] https://www.eurekalert.org/news-releases/709283

the issue is pinpointed, lowering homocysteine levels isn't difficult. And best of all, you don't need medication to do it; with compounds naturally found in food and the right supplementation, we can get the job done. Let's dive into these options.

VITAMINS B9 AND B12

One of the two pathways through which homocysteine is converted back into SAMe (see the remethylation pathways, Fig. 3a) involves vitamins B9 and B12. These pathways offer two key benefits: they help eliminate excess homocysteine and boost SAMe production. If a deficiency in either B9 (folate) or B12 is impairing efficient homocysteine recycling and supplementation is needed, the proper dosage will depend on the individual's blood levels. For vitamin B9 (folate), the optimal range is between 10 and 16 ng/mL (or µg/L), while the ideal range for vitamin B12 is between 350 and 500 pg/mL.

If supplementation is necessary, I recommend using the methylated (active) forms of these vitamins, as many people have genetic polymorphisms—especially in the MTHFR gene, which affects B9—that reduce their ability to convert the inactive form of the vitamin (the one found in food) into its active form (the one the body actually uses). A typical daily supplement dose for vitamin B9 might range from 200 to 800 µg, and for vitamin B12, between 200 and 1,000 µg.

When it comes to vitamin B9, it's possible for its blood levels to fall within the optimal range and yet homocysteine levels remain elevated. This could be due to what we just discussed: even if there is a sufficient quantity of the vitamin, it may still be in its inactive form and, because of individual genetic factors, it might not be properly converted into its active form. In such cases, supplementation with its methylated form (5-methyltetrahydrofolate [5-MTHF]) becomes essential. This is where intuition and experience play a key role: if everything appears to be within the optimal blood range but homocysteine remains high, this is likely the issue, and we should supplement with 5-MTHF—even if B9 levels appear normal. A second possible reason could be a lack of trimethylglycine (TMG) or vitamin B6. Let's take a closer look.

TRIMETHYLGLYCINE

The second pathway for recycling homocysteine into SAMe depends on TMG, also known as betaine.

Let me explain why it's one of my favorite supplements. By increasing TMG levels, we not only help lower homocysteine (the first benefit), but we also gain two additional advantages (see Fig. 3a) that are especially valuable if we want to enhance methylation capacity by increasing SAMe levels:

1. TMG facilitates the conversion of homocysteine into more SAMe, similar to how vitamins B9 and B12 work.
2. TMG is synthesized from dietary choline, meaning that supplementing with TMG spares choline from being used to produce TMG, making more of it available for phosphatidylcholine synthesis via the Kennedy pathway—a process that does not consume SAMe. As a result, more SAMe is available for other functions.

A fourth benefit is that this also increases the availability of choline for the synthesis of acetylcholine, a neurotransmitter essential for memory, learning, focus, and muscle control. Supporting acetylcholine levels is especially important for preventing or improving dementias like Alzheimer's, as we discussed earlier.

In addition to the above, TMG helps regulate fluid balance within cells, maintaining cellular volume stability and integrity under stress conditions. It also contributes to the prevention of neurodegenerative diseases and cancer, protects the liver by preventing fat accumulation, has anti-inflammatory effects, improves type 2 diabetes (by increasing insulin sensitivity and enhancing glucose homeostasis), offers cardiovascular benefits, and supports both the microbiota and the function of the intestinal and renal epithelium.[263]

In athletes, TMG helps maintain cellular volume and function during intense physical exercise.

---

[263] https://doi.org/10.1016/j.biopha.2022.112946

If taken as a supplement, I recommend the powdered form (as it's more affordable). The typical dose ranges from 3 to 6 grams per day (0.11 to 0.22 oz), or 0.05-0.1 g of TMG per kilogram of body weight (approximately 0.023-0.045 g/lb), depending on the degree of hypomethylation and homocysteine levels.[264]

As for dietary sources, TMG is found in beets, spinach, quinoa, and in the germ and bran of wheat and rye.

VITAMIN B6

The third pathway for eliminating homocysteine, known as the transsulfuration pathway (see Fig. 3b), is not used to produce more SAMe, but rather to synthesize the amino acid cysteine. Why do we need cysteine? In addition to its roles in immune function, skin and hair health, and as an expectorant for respiratory support, cysteine is essential because it is the precursor to two critical antioxidant molecules: glutathione and taurine.

Cysteine, together with glycine, is the rate-limiting factor for the production of glutathione, the most powerful intracellular antioxidant. Glutathione is one of the most important molecules in the fight against aging and in protecting the central nervous system. Beyond its antioxidant role, glutathione enhances immune response (which declines with age), regulates DNA synthesis and repair, stabilizes cell membranes, and detoxifies the body from xenobiotics—chemical compounds (some of them carcinogenic) that are not naturally produced or present in the body—such as pharmaceuticals, food additives, environmental pollutants, and industrial chemicals.[265]

As an added benefit, glutathione slows skin aging by reducing wrinkles and improving skin elasticity.[266] To carry out its function, glutathione requires not only cysteine and glycine but also the family of enzymes known as glutathione peroxidases (GPx), which depend on selenium to function effectively. Because of this, and given sele-

---

[264] https://doi.org/10.1016/j.jcm.2012.11.001
[265] https://doi.org/10.3390/molecules27010324
[266] https://doi.org/10.2147/CCID.S128339

nium's importance in thyroid function and other vital processes, it's essential to check selenium levels through blood tests and keep them within the optimal range—a goal easily achieved by eating Brazil nuts, as I discussed in my previous book, *Evolutionary Nutrition*, and will revisit in more detail in the next chapter.

On the other hand, cysteine also enables the synthesis of another amino acid with antioxidant properties: taurine. Taurine levels decline with age, so maintaining them is important. Taurine not only increases life expectancy and extends the period of healthy lifespan—improving the function of bones, muscles, pancreas, brain, adipose tissue, the gut, and the immune and cardiovascular systems—[267] but also helps counter specific hallmarks of aging. It reduces cellular senescence, protects against telomerase deficiency, suppresses mitochondrial dysfunction, decreases DNA damage, and attenuates inflammation.[268]

So far, everything seems straightforward, but it is crucial to understand that the proper functioning of this transsulfuration pathway, which produces cysteine, depends on the enzyme cystathionine beta-synthase (CBS), which in turn requires the active form of vitamin B6 (pyridoxal-5-phosphate [P-5-P]) to operate. Without this vitamin, homocysteine cannot be converted into cysteine (or into glutathione or taurine). Moreover, this pathway becomes primarily active when there is no SAMe deficiency, which means that in a context of hypomethylation, it's essential to ensure sufficient levels of SAMe (to activate the pathway) and vitamin B6 (to enable it to function properly). This will allow for the production of adequate amounts of glutathione and taurine—antioxidants that play a vital role in slowing aging, combating age-related diseases, and reducing oxidative stress.

An additional benefit of activating this pathway is that, as a byproduct of CBS enzyme activity when converting homocysteine into cysteine, hydrogen sulfide (H_2S) is produced (see Fig. 3b). H_2S is a gasotransmitter with vasodilatory effects that relaxes blood vessels and improves circulation by promoting the endothelial-derived hyperpolarizing factor (EDHF), as discussed in relation to cardiovascular

---

[267] https://doi.org/10.3390/nu15194236
[268] https://doi.org/10.1126/science.abn9257

diseases. Moreover, H₂S exerts neuroprotective effects on the nervous system, along with anti-inflammatory and antioxidant properties.

In the less common context of hypermethylation, the transsulfuration pathway will be the most activated route to eliminate homocysteine. When there is an excess of SAMe, the body does not prioritize converting homocysteine back into more SAMe (remethylation pathways), as there is already an excess, and instead prioritizes this alternative pathway. This has been observed in several cases, including schizophrenia and other conditions where an excess of SAMe coexists with elevated homocysteine and very low levels of cysteine and glutathione,[269] largely due to vitamin B6 deficiency.[270,271]

Therefore, in such cases, beyond aiming to reduce SAMe levels —using strategies like glycine and vitamin B3, as previously discussed—it is crucial to ensure sufficient availability of vitamin B6 to convert the toxic homocysteine into beneficial cysteine, thus achieving two goals simultaneously.

As always, before considering supplementation, it is important to check blood levels of vitamin B6. Optimal ranges are around 50-170 nmol/L (12.3-42 µg/L or ng/mL), and if supplementation with P-5-P is necessary to maintain these levels, the typical daily dose ranges between 2-7 mg (or 40-50 mg once a week).

VITAMIN B2

Although it only affects the process indirectly, I am including vitamin B2, or riboflavin, in this chapter because it is essential for converting inactive vitamin B9 (the folate found in foods) into its active form (see Fig. 3a). As we have seen, this conversion is crucial for recycling homocysteine.

Fortunately, vitamin B2 deficiency is rare, as it is found in a wide variety of common foods, such as dairy products, eggs, leafy green vegetables, whole grains, legumes, mushrooms, nuts, and seeds.

---

[269] https://doi.org/10.1002/bmc.5366
[270] https://doi.org/10.1016/j.redox.2021.102057
[271] https://doi.org/10.1038/s41398-021-01381-z

KEY POINTS

1. **Strategies to lower homocysteine**:
   - Importance of homocysteine. Regardless of the methylation status (hypo- or hyper-), elevated homocysteine is harmful and must be addressed, as it benefits no biological process.
   - Accurate diagnosis. Hyperhomocysteinemia (> 8 μmol/L) signals a genetic, dietary, or metabolic issue in the methionine-homocysteine cycle and should be addressed with specific strategies.

2. **Vitamins B9 and B12**:
   - Remethylation of homocysteine. Both vitamins are involved in the conversion of homocysteine into methionine and SAMe.
   - Supplementation and optimal ranges. Use the active forms of these vitamins (5-MTHF for B9 and methylcobalamin for B12) if homocysteine levels remain high despite adequate B9/B12 levels. Common dosages: B9, 200-800 μg; B12, 200-1000 μg.

3. **Trimethylglycine**:
   - Recycling and multiple effects. TMG reduces homocysteine and facilitates the synthesis of SAMe while freeing choline for other important functions, such as the production of phosphatidylcholine and acetylcholine.
   - Recommended dosage. 3-6 g daily as a powder supplement or 0.05-0.1 g/lb of body weight.

4. **Vitamin B6 and the transsulfuration pathway**:
   - Conversion of homocysteine to cysteine. Vitamin B6 activates the CBS enzyme, necessary to convert homocysteine into cysteine.
   - Additional benefits of cysteine. Increases antioxidants such as glutathione and taurine and promotes the production of H2S, which has vasodilatory and neuroprotective properties.
   - Ranges and supplementation. Optimal blood levels of B6: 50-170 nmol/L. Daily dosage of P-5-P: 2-7 mg.

5. **Vitamin B2**:
   - Role in B9 activation. Vitamin B2 converts the inactive dietary folate into its active form, essential for homocysteine recycling.
   - Adequate intake. B2 deficiency is rare, as the vitamin is found in foods such as dairy products, eggs, leafy green vegetables, whole grains, and nuts.

CHAPTER 11
How to Inhibit Gene Silencing (Deacetylation) of Tumor Suppressor Genes and Others

As you may recall from the first chapter, in addition to DNA methylation dependent on S-adenosylmethionine (SAMe), there is another key epigenetic process: histone modification. Histones are the proteins around which DNA coils within the cell. This complementary mechanism to methylation involves two main actions: acetylation (adding an acetyl group to the histone, which loosens the DNA coil, making that region more accessible and activating it) and deacetylation (removing the acetyl group, which tightens the DNA coil, compacting it and silencing that area). To carry out deacetylation, the body uses enzymes called histone deacetylases (HDACs), which are responsible for removing the acetyl group and causing the DNA to recoil, thereby reducing the expression of certain genes.

It is important to highlight that deacetylation is essential for regulating normal functions such as cell differentiation, DNA repair, and the modulation of inflammatory responses. However, in certain diseases, HDACs can become overly active and "shut down" or reduce the activity of critical genes, including tumor suppressor genes, the klotho gene (which enhances cognition and slows aging), and others related to cardiovascular disease, dementia, and metabolic disorders like diabetes.

The good news is that, in recent years, researchers have discovered specific molecules known as HDAC inhibitors (HDACi) that can help restore balance by inhibiting the excessive activity of these enzymes. This inhibition allows important genes to be reactivated and their expression to be increased. In the following sections, we will explore these molecules and their mechanisms of action, gaining a better understanding of how they can contribute to restoring a healthy epigenetic balance.

ALPHA-KETOGLUTARATE

Alpha-ketoglutarate (α-KG) is a molecule that cells use in the Krebs cycle, a key process in energy production. But beyond its role in generating energy, α-KG plays a crucial part in regulating epigenetic activity. It acts as a cofactor in the demethylation—or "reactivation"—of certain genes, such as tumor suppressor genes, as observed in non-Hodgkin lymphoma.[272]

Additionally, α-KG supports cellular self-renewal, meaning the ability of cells to remain in a primitive, versatile state, like embryonic stem cells.[273] This has potential benefits in various types of cancer, particularly those involving cancer stem cells, such as brain tumors (gliomas and glioblastomas) and leukemia, where genetic mutations often lead to decreased levels of α-KG.[274]

In short, α-KG helps sustain the activity of genes that protect against cancer, stabilizing the genome, keeping cells healthy and regulated, and preventing the uncontrolled proliferation typical of tumors.

α-KG is not found directly in foods in its pure form, but it is produced within cells from precursors found in certain foods, especially protein-rich sources, as well as tomatoes, mushrooms, pumpkin seeds, spinach, and kale.

There is also the option to supplement with calcium alpha-ketoglutarate (CaAKG) at daily doses of 2,000-4,000 mg, divided into two, three or four servings.

CURCUMIN

Curcumin is a natural compound found in the root of turmeric, a traditional spice in Indian cuisine and the main ingredient in curry.

---

[272] https://doi.org/10.1038/s41420-023-01475-1#Sec10
[273] https://doi.org/10.1038/nature13981
[274] https://doi.org/10.1016/j.ccr.2010.12.014

In addition to its well-known anti-inflammatory and antioxidant properties, curcumin also plays a key role in gene activation by inhibiting HDACs,[275] helping to keep genes that should remain active "switched on."

Besides using turmeric as a spice in everyday cooking, there are supplements that extract curcumin, the active ingredient we are interested in. If you decide to supplement, I recommend the phytosome form of curcumin (from the patented brand Meriva®), as it significantly enhances curcumin's bioavailability—up to 29 times higher compared to standard formulations (standardized extracts). This allows for greater tissue accumulation and improved effectiveness. The usual dosage is one 500 mg capsule, taken twice a day. If you opt for standard curcumin, it is best to combine it with black pepper, since piperine—the active compound in black pepper—boosts intestinal absorption of curcumin by inhibiting certain liver enzymes that would otherwise metabolize it.

EPIGALLOCATECHIN-3-GALLATE (EGCG)

Epigallocatechin-3-gallate (EGCG) is a type of catechin, an antioxidant flavonoid compound found in green tea. These compounds are widely recognized for their health benefits, including reducing the risk of cardiovascular disease, dementia, and certain types of cancer. In the field of epigenetics, EGCG has also been shown to inhibit HDACs and activate genes of great interest for the prevention and treatment of diseases.[276]

As with curcumin, in addition to incorporating green tea into our daily diet (with Jeoncha, Sencha, and matcha varieties offering the highest concentrations of catechins), there are also green tea extract supplements available. However, due to EGCG's challenges with bioavailability—caused by its rapid metabolism and limited absorption—I recommend choosing the phytosome form of green tea extract, patented under the name Greenselect®. This formulation binds EGCG

[275] https://doi.org/10.3390/molecules27134014
[276] https://doi.org/10.1016/j.yexcr.2015.04.004

to phospholipids, significantly enhancing its absorption and the body's ability to utilize this compound. The usual dosage for this formulation is one 250 mg capsule, taken twice a day.

ALLICIN

Allicin is a compound formed when fresh (uncooked) garlic is cut or crushed, released as the plant cells are broken. It is responsible for garlic's characteristic smell and flavor. However, allicin is unstable and quickly breaks down into several compounds, two of which are of particular interest to us: diallyl disulfide and allyl mercaptan. These are the main contributors to garlic's health benefits, including its anticancer effects and its ability to inhibit HDAC activity—[277] giving us yet another reason to include this superfood in our diet.

Although there are supplements that provide around 5,000 μg of allicin per capsule, I do not recommend them, as it is very easy to reach that amount by simply consuming one or two cloves of garlic per day. That said, to ensure maximum allicin production, it is crucial that the garlic be fresh and that it be crushed or minced at least five minutes before consumption.

KAEMPFEROL

Kaempferol is a natural flavonoid found in a wide variety of fruits and vegetables, especially kale, spinach, broccoli, onions, apples, grapes, citrus fruits, and green tea. This compound has been extensively studied for its ability to reduce the risk of chronic diseases, particularly cancer, thanks to its antioxidant and anti-inflammatory properties. In the field of epigenetics, kaempferol has been shown to inhibit HDAC activity by up to 60%.[278]

Although there are some products that offer kaempferol as a supplement, they are quite rare, and I do not trust the quality of the

---

[277] https://doi.org/10.2174/187152011795347540
[278] https://doi.org/10.1007/978-3-319-31143-2_62-1

brands that manufacture them, so I do not recommend supplementation. The best approach is to regularly include the foods mentioned above in your diet to naturally benefit from kaempferol.

PROANTHOCYANIDINS

Proanthocyanidins are a group of polyphenols widely found in plant-based foods, especially in grape seeds, raw and pure cocoa, and berries (such as blueberries, blackberries, and strawberries), as well as apples. These compounds possess antioxidant and anti-inflammatory properties and have demonstrated beneficial effects on cardiovascular health and blood glucose regulation. In the context of epigenetics, proanthocyanidins have the ability to activate tumor suppressor genes and other protective genes.

Among proanthocyanidins, the ones found in grape seeds—specifically in their oligomeric proanthocyanidin (OPC) form—are the most studied for their epigenetic effects. This particular subtype is more potent and has greater biological activity due to its unique structure, allowing it to interact more effectively with cellular systems involved in epigenetic regulation, and to more efficiently reactivate the genes we aim to keep active.

If you choose to supplement, it is important to ensure that the supplement contains the OPC form, rather than the more common "standard" grape seed extract. The typical recommended dosage is one to two capsules per day (taken separately if taking two), making sure each capsule contains around 100 mg of OPC.

GENISTEIN

Genistein is a type of isoflavone primarily found in soybeans and their derivatives, such as tofu and tempeh. Due to its chemical structure, which is similar to that of human estrogens, genistein can bind to estrogen receptors, with important implications for hormonal health and the prevention of certain types of cancer, as it also has the ability to inhibit HDACs. This inhibition helps activate genes that can suppress tumor growth and induce programmed cell death in cancer cells.

In addition to its epigenetic benefits, genistein also possesses anti-inflammatory and antioxidant properties.

It is important to highlight that breast cancer deserves special consideration when discussing genistein. This compound can exhibit both estrogenic and anti-estrogenic properties, meaning it can mimic estrogen in some circumstances and block its effects in others, depending on which estrogen receptor (ER) it binds to (ER-α or ER-β). Genistein has a greater affinity for the ER-β receptor, where it acts as an agonist, enhancing its activity. Conversely, when it interacts with the ER-α receptor, it acts as an antagonist, suppressing its expression and activity. According to some studies, both mechanisms contribute to genistein's anticancer activity in breast cancer.[279, 280]

However, these effects have only been observed at blood concentrations of genistein higher than 5 μM (5 micromoles of genistein per liter of plasma) after supplementation or food absorption. Achieving this concentration through dietary soy intake or typical genistein (or isoflavone) supplementation is nearly impossible, as it would require consuming around 4.5 g of genistein,[281] whereas typical supplement doses are about 100 mg per day.

On the other hand, lower concentrations of genistein might actually stimulate the growth of estrogen receptor alpha-positive (ER-α+) breast cancer cells, which are hormone-sensitive and represent the most common type of breast cancer.[282]

Given that the genistein dosage necessary to ensure benefits in ER-α+ breast cancers is not clearly established and appears to be quite high, I recommend applying the "precautionary principle": avoid supplementing with genistein or soy isoflavones and consult with an oncologist before consuming soy and its derivatives in these types of cancers.

To clarify any confusion, it is important to note that soy lecithin does not contain genistein, as genistein is an isoflavonoid present in the protein portion of soy, while lecithin is extracted from the fat

---

[279] https://doi.org/10.3390/pr10020415
[280] https://doi.org/10.3390/cimb43030106
[281] https://doi.org/10.1097/gme.0b013e318167b8f2
[282] https://doi.org/10.3389/fphar.2017.00699

portion, which does not contain genistein. Therefore, due to its ability to conserve SAMe and the benefits this provides in cancer, lecithin can not only be safely consumed in cases of breast cancer but is actually advisable.

On the other hand, in breast cancer cells that do not depend on estrogen for growth (ER−), a less common type of breast cancer, genistein—regardless of dosage—shows an antiproliferative and beneficial effect by inhibiting the growth of cancer cells.

As I mentioned earlier regarding soy lecithin to increase phosphatidylcholine and conserve SAMe, when consuming soy and its derivatives, it is important to ensure that they carry the "IP" and "Non-GMO" labels. The IP certification guarantees that the soy product has been tracked and kept separate throughout the entire production chain—from seed to final product—ensuring that the specific characteristics of the soy variety and cultivation method are preserved.

The Non-GMO label indicates that the product has not been genetically modified, which could otherwise raise potential concerns regarding allergies, antibiotic resistance, and other possible long-term effects.

Genistein supplementation can be found in three forms:

1. as soy extract (approximately 40% is isoflavones and 15% is genistein);
2. as soy isoflavones (about 38% of which is genistein); and
3. as pure genistein.

In any case, what matters is the actual genistein content, and any supplement should clearly indicate this amount per capsule on the label. The typical daily recommended dosage ranges between 90 and 180 mg; if opting for the higher end of this range, it is best to divide it into two separate doses. Genistein can also be derived from the plant *Sophora japonica* (the Japanese pagoda tree) instead of soy; in that case, it is not necessary to look for the Non-GMO certification. Nonetheless, I must emphasize once again: genistein supplementation should be avoided in estrogen receptor-positive (ER+) breast cancers.

RESVERATROL

Resveratrol is a stilbene, a type of natural polyphenol found in various plants, especially in red grapes. It is widely recognized for its antioxidant and anti-inflammatory properties. In the context of epigenetics, resveratrol has also been shown to inhibit HDACs, which, as we have seen, facilitates the expression of genes that can suppress tumor growth and promote programmed cell death in cancer cells.[283]

However, one significant challenge with resveratrol is its low bioavailability; the body metabolizes it rapidly before it can reach the cells in effective amounts. The doses considered effective for these epigenetic benefits in humans are as high as 1 g per day—a quantity virtually impossible to achieve through grape consumption alone, making supplementation the only practical way to reach these levels.

There is a more effective alternative for supplementation: trans-pterostilbene, an active (methylated) form of resveratrol with much higher bioavailability. Thanks to its improved absorption, the required dosage is approximately ten times lower; a daily dose of 100 mg is recommended, divided into two separate 50 mg servings.

BUTYRATE

Butyrate is a type of short-chain fatty acid—one of the simplest fat molecules—that plays a fundamental role in gut health and in other systems of the body. Butyrate is primarily produced in the colon (large intestine) when beneficial bacteria break down (ferment) dietary fibers that the human body cannot digest. This molecule is fascinating, and we should make an effort to boost its production because of its wide range of benefits. Among its functions, butyrate reduces inflammation and modulates the immune system, increases the expression of antimicrobial proteins, enhances gut health by serving as an energy source for the cells lining the intestine, acts as a neuroprotective agent, and inhibits the growth of cancer cells by blocking HDACs, among other mechanisms.

---

[283] https://doi.org/10.1371/journal.pone.0073097

One of the most effective ways to increase butyrate production is by boosting the intake of resistant starch. Resistant starch is a type of prebiotic carbohydrate that is not digested in the small intestine, meaning it reaches the colon intact, where it is fermented by bacteria to produce butyrate. The best way to increase resistant starch content in the diet is by cooking oats, rice, or potatoes the day before consumption and then refrigerating them after they have cooled to room temperature. This process significantly raises the amount of resistant starch in these foods. Green plantains are also an excellent source of resistant starch. For those who are not fond of their taste, green plantain flour can be used instead, added to any dish.

There are additional strategies to encourage butyrate production by gut bacteria. Increasing the intake of soluble and insoluble fiber —found in fruits, vegetables, nuts, seeds, legumes, and whole grains— is one of them. Pectin, a particularly interesting type of soluble fiber for boosting butyrate, is mainly found in the skin of green apples and in citrus fruits (oranges, lemons, grapefruits), especially concentrated in the white part of the peel. To maximize pectin intake, it is advisable to peel citrus fruits in a way that preserves the white inner layer. Apple fiber powder is another convenient option to enhance pectin intake in meals.

In addition to resistant starch and pectin, other prebiotics that promote butyrate production include chicory, garlic, onions, leeks, artichokes, cabbages, and asparagus. However, feeding butyrate-producing bacteria won't be very effective if these bacteria are not sufficiently present. Therefore, it is also important to increase the intake of probiotics (butyrate-producing bacteria), found in foods such as yogurt, kefir, sauerkraut, kombucha, kimchi, miso, tempeh, and nattō.

Lastly, we must not forget that regular physical exercise has also been shown to boost butyrate production by gut bacteria, providing yet another powerful benefit of staying physically active.

CHLOROGENIC AND CAFFEIC ACIDS

Chlorogenic and caffeic acids are two phenolic compounds found primarily in coffee, although they are also present in smaller amounts in tea, cocoa, apples, blueberries, pears, and tomatoes. These compounds offer important health benefits, including antioxidant, anti-inflammatory, and anticancer effects. Their ability to inhibit HDACs makes them potentially valuable agents for the prevention and treatment of chronic diseases and cancers. Yet another benefit of coffee.

SELENIUM

Selenium is an essential mineral that plays a crucial role in regulating the immune system and possesses powerful antioxidant properties, as it is necessary for glutathione to function effectively, as we discussed in the previous chapter. In addition to these benefits, recent research has revealed that selenium also influences epigenetics by inhibiting HDACs, giving it the potential to slow cancer progression.

As I mentioned earlier, selenium blood levels are rarely checked in routine lab work, even in cancer patients. However, in my practice, it is a fundamental biomarker that I always include in my panels. According to studies, the optimal blood selenium range for reducing the risk of various diseases is between 130-135 ng/ml (or µg/l).[284, 285] When we speak specifically about cancer risk, the optimal level appears to be even slightly higher—around 150 ng/ml—according to a meta-analysis conducted by a Chinese university.[286]

The richest dietary source of selenium is Brazil nuts. Whenever possible, I prefer using food rather than supplements to achieve optimal nutrient levels, and in the case of selenium, it is relatively easy to do so with Brazil nuts. To reach a blood level of 130 ng/ml, consuming about two to three Brazil nuts per day should be sufficient, assuming an average selenium content of 70-90 µg per nut, which is typical. This range provides a daily intake of approximately 140-270 µg of selenium, which is adequate to maintain healthy blood levels without exceeding toxicity thresholds.

---

[284] https://doi.org/10.3389/fcvm.2021.671618

[285] https://jamanetwork.com/journals/jamainternalmedicine/fullarticle/414000#:~:text=At serum selenium levels of,130 to 150 ng/mL

[286] https://doi.org/10.1038/srep19213#Sec21

That said, it is always essential to first check each individual's current selenium levels because, as with almost everything in the body, both deficiency and excess can be harmful. At excessively high levels, selenium can lead to toxicity, with symptoms including hair loss, gastrointestinal problems, and neurological damage.

QUERCETIN

Quercetin is a natural flavonoid found in a wide variety of fruits (such as citrus fruits, apples, red grapes, and berries) as well as vegetables (including onions, broccoli, kale, garlic, leeks, and peppers). It is also present in green tea, cocoa, and other foods. Quercetin is well known for its antioxidant, anti-inflammatory, and anticancer properties; in recent years, it has gained attention for its ability to inhibit HDACs and facilitate the activation of tumor suppressor genes.

If supplementation is chosen, the typical recommended dosage is around 500 mg of quercetin per day.

A word of caution. Quercetin inhibits the activity of the enzyme acetaldehyde dehydrogenase (ALDH), which, during alcohol metabolism, is responsible for converting acetaldehyde (a toxic compound that causes headaches) into acetate (a harmless substance). Therefore, if you are taking quercetin supplements or follow a diet high in foods containing quercetin, be aware that this can further amplify the negative effects of alcohol.

SULFORAPHANE

Sulforaphane is a natural organic compound belonging to the isothiocyanate family, primarily found in cruciferous vegetables such as broccoli sprouts, Brussels sprouts, kale, cauliflower, and cabbage. It is renowned for its powerful antioxidant, anti-inflammatory, and anticancer properties, and it stands out particularly for its ability to inhibit HDACs and modulate other epigenetic pathways.

Regarding supplementation, broccoli seed extracts are available that contain approximately 30 mg of sulforaphane per capsule, with up to two capsules per day being a common dosage.

Moreover, the combined use of sulforaphane and EGCG has been shown to produce a synergistic effect, enhancing each other's ability to reactivate tumor suppressor and protective genes such as *p21* and *klotho*.[287]

The *p21* gene is a major tumor suppressor that regulates the cell cycle, and increasing its expression helps to halt cell cycle progression and allow DNA repair, thereby preventing the proliferation of cancer cells.

On the other hand, the *klotho* gene is well known for its anti-aging and protective effects across various tissues. It improves insulin sensitivity, reduces oxidative stress, and provides neuroprotective and cognitive benefits, making it a crucial target for health and longevity. It is not surprising that physical exercise is one of the most powerful known enhancers of the *klotho* gene, as demonstrated by a research team at the University of Granada.[288]

Additionally, it has been shown, specifically in breast cancer, that vitamin D can enhance the anticancer effects of sulforaphane.[289]

VITAMIN D

I will not go into great depth about this vitamin (hormone) here because I have already covered it extensively in my previous book, *Alimentación evolutiva*, as well as in a detailed article[290] on my blog, where I explain its nearly infinite functions, its relationship to various diseases, optimal levels, types, sources, supplementation, and more.

Regarding its connection to the topic of this book, vitamin D does not act as a direct HDAC inhibitor, unlike some of the compounds we discussed earlier. Its epigenetic action is different: instead of directly inhibiting HDACs, vitamin D indirectly influences their activity through the activation of the vitamin D receptor (VDR), which facil-

---

[287] https://doi.org/10.1016/j.mce.2015.02.020
[288] https://doi.org/10.1080/02640414.2019.1626048
[289] https://doi.org/10.1111/jfbc.13114
[290] https://www.curroclavero.com/vitamina-d-version-extendida-parte-1-de-2-hormona-ciclo-tipos-funciones-y-relacion-con-enfermedades

itates the expression of tumor suppressor genes and enhances the effectiveness of HDAC inhibitors.[291]

Interestingly, this relationship also works in the opposite direction: the HDAC inhibitors we discussed earlier also enhance the effect of vitamin D by increasing the expression and activity of its receptor.[292] These epigenetic changes, induced by the synergistic interaction between HDAC inhibitors and vitamin D, improve the anticancer effects of both agents individually, creating a favorable environment for the expression of protective genes.

Is there another dietary molecule that acts synergistically with vitamin D?

Yes—omega-3 fatty acids, which, when combined with vitamin D, have been shown to reduce the incidence of advanced (metastatic or fatal) cancer, particularly in individuals who are not overweight.[293] It is important to keep this last factor in mind.

OMEGA-3 FATTY ACIDS

I have also discussed omega-3 fatty acids extensively in my previous book and on my blog.[294] These fatty acids are well known for their multiple health benefits, including anti-inflammatory, cardioprotective, and neuroprotective effects. However, what is less widely known is that omega-3s may also play a crucial role in epigenetic regulation and cancer inhibition by reducing the expression of EZH2,[295] an enzyme that contributes to gene silencing through methylation. This epigenetic change can reactivate tumor suppressor genes that had previously been silenced by this enzyme,[296] providing a pathway to halt the proliferation of cancer cells and to promote apoptosis, a programmed cell death process essential for eliminating cancerous cells.

---

[291] https://doi.org/10.1007/s13148-011-0021-y
[292] https://doi.org/10.3390/app10124096
[293] https://doi.org/10.1001/jamanetworkopen.2020.25850
[294] https://www.curroclavero.com/omega-3-papel-crucial-en-proceso-desinflamacion-otros-beneficios-necesidad-diaria-fuentes-naturales-y-suplementacion
[295] https://doi.org/10.1093/carcin/bgp305
[296] https://doi.org/10.1186/s13148-021-01034-4

***

Before concluding this chapter, I want to highlight an important point: although specific supplements exist for many of the compounds mentioned above, I do not recommend supplementing with them as a preventive measure against disease—and certainly not combining multiple supplements at the same time. If there is a particular pathology or a predisposition to one, it would be necessary to carefully evaluate and determine which of these options are the most appropriate.

Nevertheless, I do encourage regularly incorporating foods rich in these compounds into our daily diet. These foods include: broccoli and broccoli sprouts, kale, garlic, red grapes, onions, cabbages, apples, berries, soy and its derivatives (tofu, tempeh), Brazil nuts, turmeric, oily fish (such as sardines, mackerel, tuna, and salmon), carrots, spinach, citrus fruits (oranges, grapefruits, lemons), nuts and seeds, oats, green tea (Sencha and matcha varieties), coffee, and cocoa (pure raw powder or nibs), among others.

The importance of consuming these foods in their whole form lies in a fundamental concept: the *food matrix*—the natural structure and organization of all nutrients and bioactive compounds within a whole food. This matrix has a direct impact on how these nutrients are absorbed and utilized by the body, as the way they are organized affects their bioavailability and health effects.

For example, an orange provides not only vitamin C but also fiber, water, and other compounds that are naturally arranged in a way that enhances absorption and modulates the body's response. Eating a whole orange is not the same as drinking only its juice or taking a vitamin C supplement: juice lacks the original fiber and structure, and supplements do not offer the food matrix that only the whole fruit can provide.

On the other hand, as you can see, the list of foods rich in molecules that activate antitumor and other beneficial genes does not include products like sausages, cookies, industrial bread, cold cuts, hamburgers, pizza, breaded meats, pasta, breakfast cereals, pastries, cakes, or ice cream. These edible products (which are not truly foods) not only lack scientific evidence supporting their ability to prevent disease, but, in fact, the evidence shows precisely the opposite: regular consumption of these items is harmful to health.

It is nature—and not the food industry—that provides us, through its whole foods, with the molecules that help prevent and improve the prognosis of the chronic diseases that concern us so much and cause so many premature deaths. Yet we continue to base our diets on processed products made by the food industry, whose main priority, like that of any other industry, is to sell—not to protect or care for our health. I'm not saying that these products cannot have an occasional place in our diet, especially for people who are healthy and maintain good habits. However, they must not form the foundation of our nutrition.

It deeply saddens me every time I go grocery shopping and see what people are putting into their carts—products they will eventually consume. I see their levels of overweight and inflammation, which drastically increase the risk of disease and accelerate aging. I also observe people who would rather wait two minutes for the elevator than walk up a single flight of stairs, and others who claim they don't have time to exercise yet spend more than three hours a day in front of screens: social media, frivolous news (often false or manipulated), absurd videos, unfunny memes, and other fleeting distractions. It is no surprise that hospitals and healthcare systems are overwhelmed and collapsing, and that despite the tremendous advances in science and medicine, so many preventable diseases and premature deaths continue to occur.

I firmly believe that simply adopting a diet based on the foods provided by nature and avoiding a sedentary lifestyle would account for 90% of the benefits in terms of physical health and quality of life. From there, we can take additional steps to aspire to the *cum laude* level, which is one of the goals of this book. But the above is the foundation on which everything else must be built. This is the message that I believe nutrition professionals should focus on, rather than getting caught up in pointless debates about which diet is the best among the countless ones that exist, or whether drinking lemon water on an empty stomach is beneficial, or if fruit should be eaten before or after a meal, among so many other trivialities.

KEY POINTS

1. **Importance of Histone Acetylation and Deacetylation**:
   - Epigenetic modification of histones: Acetylation loosens DNA, activating gene expression, while deacetylation compacts DNA and silences genes. Deacetylation is regulated by HDAC enzymes.
   - HDACs and disease: Excessive HDAC activity can silence tumor suppressor genes and other protective genes, contributing to diseases such as cancer, diabetes, and dementia.
   - HDAC inhibitors (HDACi): Natural molecules that can inhibit HDACs and reactivate beneficial genes when needed.

2. **Molecules That Help Slow Down Deacetylation**:
   - α-KG: Participates in the Krebs cycle and regulates epigenetic activity. Found in protein-rich foods and vegetables such as spinach, tomatoes, and mushrooms.
   - Curcumin: The active compound in turmeric. Inhibits HDACs. As a supplement, the phytosome form of curcumin (Meriva®) enhances bioavailability. If using standard curcumin, it should be combined with black pepper to improve absorption.
   - EGCG: A catechin from green tea that inhibits HDACs and activates protective genes. In supplement form, the phytosome extract of green tea (Greenselect®) improves absorption.
   - Allicin: A bioactive compound from garlic. Allicin and its derivatives inhibit HDACs; ideally consumed by eating 1-2 cloves of fresh crushed garlic daily.
   - Kaempferol: A flavonoid found in kale, spinach, and fruits that reduces HDAC activity. It is best obtained through the diet, as supplement quality is unreliable.
   - Proanthocyanidins: Polyphenols from grape seeds (OPCs), cocoa, and berries that activate tumor suppressor genes.
   - Genistein: A soy isoflavone. Inhibits HDACs and may have hormonal effects beneficial in certain non-hormonal cancers (ER−). Caution is needed in ER+ breast cancer; consultation with an oncologist is recommended before supplementation.

- Resveratrol: An antioxidant and anti-inflammatory polyphenol found in red grapes. It has low bioavailability; as a supplement, trans-pterostilbene is a more effective alternative.
- Butyrate: A short-chain fatty acid produced by the gut microbiota. Inhibits HDACs and improves intestinal health. The best way to enhance butyrate production is by increasing resistant starch and fiber intake, especially pectin.
- Chlorogenic and caffeic acids: Found in coffee. Inhibit HDACs and possess antioxidant and anticancer properties.
- Selenium: An antioxidant mineral especially abundant in Brazil nuts. Inhibits HDACs and enhances glutathione activity.
- Quercetin: An antioxidant flavonoid present in various fruits and vegetables. Inhibits HDACs.
- Sulforaphane: Found in cruciferous vegetables. Inhibits HDACs and boosts the expression of protective genes such as *p21* (tumor suppressor) and *klotho* (anti-aging).
- Vitamin D: Activates the VDR receptor, promoting the expression of tumor suppressor genes and synergizing with HDAC inhibitors. Combined with omega-3s, it enhances the reduction of advanced cancer risk.
- Omega-3: Reduces EZH2 expression, allowing the reactivation of tumor suppressor genes.

3. **Conclusion**:
    - Encourage whole-food consumption: Build a diet rich in natural foods such as cruciferous vegetables, garlic, fruits, berries, green tea, cocoa, and oily fish.
    - Avoid preventive supplementation: It is better to consume foods containing bioactive molecules in their natural food matrix rather than rely on multiple supplements without professional guidance.

EPILOGUE

Throughout this book, we have journeyed through the emerging world of nutritional epigenetics, a field that is revolutionizing our understanding of how nutrients can shape biology in unexpected ways. What we once considered simply as "food" we now recognize as a powerful tool capable of influencing gene expression and, as a result, impacting health and longevity.

We have explored how research in this new field is expanding our knowledge of the ways nutrients and other dietary components can influence gene expression. These discoveries are opening new possibilities for preventing and treating chronic diseases, promoting health, and slowing the aging process through much more personalized nutrition.

Thanks to new tools and knowledge, we are moving toward a form of personalized nutrition that allows us, with increasing precision, to optimize individual health by adjusting foods and their molecules according to each person's genetic and, more importantly, epigenetic characteristics. After decades of intense research, we have reached a point where our perspective on food has radically changed. We no longer see food merely as something that satisfies hunger and provides energy and nutrients; we now recognize it as "functional food" that influences key epigenetic processes within cells, offering specific health benefits.

In other words, we are transforming foods (and the molecules they contain) into "medicines" that, depending on each person's state, can be directed toward preventing and treating disease, as well as providing the body with the tools it loses with age—tools necessary to slow down the global deterioration associated with aging.

Throughout this book, I have spoken extensively about S-adenosylmethionine—"SAMe" to its friends—my "pride and joy" among the many molecules circulating in the body. I hope that everything you have read so far has helped you understand the importance of this molecule, which remains largely unknown to most people, and the reasons behind my passion for it ever since I first learned of its existence: without it, we could neither survive nor adapt to the world around us, and part of the global deterioration the body experiences with age is due to the loss of SAMe over time.

A great deal of information has been presented throughout this book, and I believe now is the right moment to recap.

We can increase the availability of SAMe through two key strategies:

1. by conserving SAMe, supplying externally the molecules that consume the most SAMe during their synthesis (such as creatine and lecithin/phosphatidylcholine); and
2. by converting homocysteine into more SAMe.

By applying these two strategies, we achieve five vitally important health and longevity benefits:

1. **Supporting DNA methylation**. This is one of the two main processes of epigenetics, and it deteriorates with age. This decline is associated with DNA instability and numerous diseases and is one of the primary drivers of aging.
2. **Synthesizing spermidine**. We have discussed how important spermidine is for longevity and health. Although it can be obtained through certain foods, the body can also synthesize it endogenously using SAMe.
3. **Supporting the GNMT pathway**. This pathway activates to recycle excess SAMe via glycine. It is crucial for aging for two reasons: first, overexpression of this pathway extends lifespan; second, without SAMe recycling, the lifespan extension associated with the polyamine spermidine does not occur.
4. **Increasing glutathione levels**. Glutathione is the most powerful intracellular antioxidant we have. With an adequate supply of SAMe, there is less need to recycle homocysteine back into SAMe, which favors the transsulfuration pathway.

With sufficient vitamin B6, this pathway converts homocysteine into cysteine for glutathione production.
5. **Increasing taurine levels**. Similarly, the cysteine resulting from the removal of homocysteine serves as a precursor for taurine. This extracellular antioxidant not only extends lifespan but also improves the functioning of several key systems in the body.

We have also explored why lowering homocysteine levels is beneficial: not only because homocysteine can be converted into antioxidants like glutathione and taurine (or into more SAMe if needed), but also because simply reducing its levels is positive in itself, given its toxic role and its association with an increased risk of cardiovascular disease, dementia, and many other conditions.

I must emphasize that it is crucial for all bodily processes to function properly. In my opinion, however, alongside the Krebs cycle—which produces energy from food—there is no process more important to overall health than the proper functioning of the methionine-homocysteine cycle and the compensation for SAMe loss with aging.

But beyond the science, what truly matters is how we apply this knowledge in our daily lives. Nutritional epigenetics provides us with a roadmap for living a healthier and fuller life—but it is up to us to take the helm.

I hope the theoretical part of this book has served as a guide to understanding the foundations behind our biology and the countless processes the body carries out without our awareness, all aimed at helping us survive and adapt as effectively as possible to the ever-changing world around us.

I also hope the practical part has provided valuable strategies for making changes that will help you reach your fullest potential, slow down your aging process, and reduce the risk of diseases that could lead to premature death or a significant decline in quality of life.

We are standing at the threshold of an era where nutritional epigenetics and personalized diets are becoming key tools for optimizing health throughout life. We only have one life in this world, and it is a miracle that we are here—so let's make the most of this opportunity, extend our lives a little longer, and, most importantly, strive to live as dignified a life as possible.

To conclude, I want to say that although everything presented in this book is based on science, fortunately, science never stands still —and especially in this field, it is growing at a tremendous pace. The science of nutritional epigenetics will continue to evolve, and with it, our possibilities for living better lives. We will keep learning and expanding our understanding over time. For that reason, I will continue sharing new information on my blog, which you can find on my website (www.curroclavero.com/en), where you can also subscribe to my newsletter to receive email notifications whenever I publish a new article or book.

Finally, if you feel that the content of this book has been worthwhile, I would sincerely appreciate a review on Amazon, as it is one of the best ways to help it reach more people.

This book is just the beginning. My hope is that the ideas presented here have inspired you to take charge of your health in a new, more informed, and more intentional way. The future of nutrition is full of possibilities, and now you have the tools to make the most of them.

Thank you for accompanying me on this journey.

Knowledge is power—and now, the power is in your hands.

APPENDICES

Microbiota and Methylation

We are becoming increasingly aware of the importance of maintaining a healthy microbiota for overall health—that is, preserving a balanced and diverse community of microorganisms (bacteria, viruses, fungi, etc.) in the gut. This diversity and balance allow the different microbial species to optimally perform their specific functions, such as vitamin production, dietary fiber fermentation, protection against pathogens, and modulation of the immune system. A gut with a diverse microbiota is more resilient to disruptions such as infections and helps prevent chronic diseases like obesity, type 2 diabetes, cardiovascular disease, cancer, dementia, inflammatory bowel disease, psychiatric disorders, gastrointestinal disorders, and many others. I will not delve further into this topic here, as it would take us away from the main focus of the book.

In addition to everything mentioned earlier, the latest evidence indicates that the microbiota also has the ability to carry out epigenetic modifications. In fact, it is suggested that this capacity may be the primary mechanism through which the microbiota exerts such a significant impact on health.[297] For example, a 2022 study demonstrated that a lack of a healthy microbiota reduces levels of S-adenosylmethionine (SAMe) and, consequently, diminishes global methylation capacity.[298] Similarly, it has been shown that maintaining good levels of SAMe combined with a healthy microbiota produces a synergistic effect, mitigating dysbiosis and systemic inflammation, and improving conditions such as depressive states.[299]

---

[297] https://doi.org/10.1038/s41564-019-0659-3
[298] https://doi.org/10.3389/fmicb.2022.1065668#h5
[299] https://doi.org/10.3390/nu14132751

Different strains of gut bacteria can produce SAMe from methionine, meaning that changes in bacterial composition can influence SAMe availability and alter the DNA methylation status.[300] Moreover, bacteria are also responsible for producing other molecules that can modify gene expression, such as butyrate—which, as we have already discussed, has significant effects on conditions like inflammatory bowel disease, cancer, and obesity.[301]

These epigenetic modifications induced by the microbiota, such as DNA methylation, influence the regulation of gene expression and have a particularly significant impact on immune system function. As we discussed at the beginning, the most important thing is to maintain a diverse and balanced community of microorganisms, as alterations in the composition and structure of the microbiota (dysbiosis) lead to disease.

VITAMIN B12

Although we do not yet fully understand how the microbiota influences these epigenetic changes, we are gradually uncovering more details. For example, we now know that the role of vitamin B12 is fundamental and closely interrelated with the microbiota in a bidirectional manner.

On one hand, an alteration in the microbiota, such as small intestinal bacterial overgrowth (SIBO), reduces the amount of available B12, since these bacteria—which should not be present there—consume and sequester B12 before it can be absorbed in the intestine. In other words, they compete with the human body for this vitamin.[302] Moreover, these bacteria can also convert B12 into forms that are unusable by the body. This situation is even more serious because it would not necessarily be detected in a blood test; that is, we could have B12 present, but it would not be functional. Both of these factors prevent the body from converting homocysteine into SAMe (see Fig. 3a),

---

[300] https://doi.org/10.1007/s11274-016-2102-8
[301] https://doi.org/10.1080/19490976.2021.2022407
[302] https://doi.org/10.1053/j.gastro.2020.06.090

which, as we know, reduces methylation capacity and affects proper gene expression. In such cases, high doses (> 1000 μg) of methyl-B12 (its bioactive form) taken orally as supplements or administered via subcutaneous injections (which bypass the gastrointestinal tract) can be a solution.

On the other hand, a deficiency of B12 negatively affects the composition of the intestinal microbiota[303] and increases intestinal permeability, allowing molecules or pathogens that should not enter the bloodstream to do so, thereby raising the risk of infections.[304] B12 is essential for various metabolic functions in both humans and bacteria, and its deficiency can lead to an imbalance in the gut microbial populations, affecting their diversity and functionality (dysbiosis). This, in turn, further reduces B12 availability and functionality, creating a vicious cycle.

OMEGA-3

To complicate this equation even further, we must also consider omega-3 fatty acids. Omega-3s—especially DHA—are essential for proper brain function, but to fulfill this role they require methyl groups from SAMe, as well as a healthy microbiota and adequate levels of B12. A B12 deficiency reduces brain DHA levels and disrupts proper BDNF function, as has been observed in cases of autism and attention deficit hyperactivity disorder (ADHD). Furthermore, the gut microbiota can also alter omega-3 levels, enhancing them when the microbiota is healthy and diminishing them when it is not. In turn, omega-3s can modify the bacterial species in the gut, promoting anti-inflammatory strains like *Lactobacillus* and *Bifidobacterium*, or causing dysbiosis in cases of omega-3 deficiency.

This intricate relationship between the microbiota, SAMe, vitamin B12, and omega-3s is yet another example of how the body functions as a complex system. Everything is interconnected: the whole affects the parts, and the parts affect the whole. Specific changes produce

---

[303] https://doi.org/10.1093/advances/nmab123
[304] https://doi.org/10.1186/s40168-023-01574-2#Sec11

adaptations in the entire system, and the system, in turn, influences the specifics, creating an organism in constant motion and adaptation to its environment. This is why I often say that the health sciences are more an art than a science.

All the research in this field has only just begun, and as we mentioned earlier, the complex interaction and mechanisms between the microbiota and its impact on epigenetics are still not well understood. We will need to stay attentive, but for now, what we do know for certain is that taking care of our gut bacteria is crucial—because, regardless of how they do it, one thing is clear: they take care of us.

How SAMe and Methylation Affect Physical Performance

In recent years, researchers have begun investigating how high-intensity physical exercise impacts DNA methylation—that is, what happens to the body from an epigenetic perspective when we engage in intense exercise, whether during training or competition. Early evidence has shown that both occasional and continuous exercise positively influence DNA methylation,[305,306] in a manner similar to the effects of a healthy microbiota. As we discussed in the first chapter of this book, the increase in lactate—a byproduct of energy metabolism that rises during exercise—affects one of the two major epigenetic mechanisms: histone modification. This recently discovered process is known as lactylation, and it promotes the activation of beneficial genes. But the relationship is also bidirectional: epigenetics likewise affects physical performance.

Understanding these mechanisms is vital not only for the health of the general population but also for athletes, who can leverage this knowledge to develop strategies that optimize their performance in competitions and enhance the desired adaptations following training.

In 2019, it was discovered that immediately after completing a high-intensity endurance exercise session, two things occur:

1. the expression of the gene that produces the protein PGC-1α increases, and
2. global DNA methylation temporarily decreases.[307]

---

[305] https://doi.org/10.1111/apha.12414
[306] https://doi.org/10.1080/15592294.2019.1614416
[307] https://doi.org/10.1080/15592294.2019.1582276

Let's take a closer look at both phenomena.

INCREASE IN PGC-1α

First, regarding the increase in PGC-1α, I will not dwell too much on it, as it is a well-known phenomenon. PGC-1α is a protein that plays a crucial role in four main processes:

1. **Regulation of energy metabolism**: It stimulates fatty acid oxidation, facilitating the use of fats as an energy source. It also enhances the uptake and utilization of glucose by muscle cells.
2. **Mitochondrial biogenesis**: It not only assists in the creation of new mitochondria—the cellular "engines" where most energy is produced—but also improves the efficiency of existing mitochondria.
3. **Inflammation and oxidative stress**: PGC-1α has anti-inflammatory effects that help reduce chronic inflammation and enhance the cell's antioxidant defenses, protecting it against oxidative damage. Both functions contribute to reducing cellular aging and lowering the risk of chronic diseases.
4. **Muscular adaptation to exercise**: It improves aerobic capacity and endurance, allowing prolonged effort with less fatigue.

In summary, the increase in this protein triggered by exercise is one of the many mechanisms that make training so beneficial for systemic health.

SHORT-TERM REDUCTION (AND LONG-TERM INCREASE) IN GLOBAL METHYLATION

Now we will address the second phenomenon, which, besides being the most novel, is more directly related to the theme of this book. It has been shown that immediately after completing a high-intensity physical activity session, there is a short-term decrease in the body's global DNA methylation capacity. Among the family of DNMT enzymes—those involved in DNA methylation—DNMT3a and

DNMT3b, which are the most dependent on SAMe, experience a reduction in their expression. This suggests that during high-intensity exercise, there is such a high demand and consumption of SAMe that the body's methylation capacity becomes depleted following the activity.

Continuing along this line of research, in 2020, the University of Oslo conducted another study[308] that confirmed and expanded upon previous findings. After analyzing blood samples from a group of well-trained athletes before and after a session of high-intensity physical activity, a significant decline in their methylation capacity was again detected. This time, the researchers used a different metric: the methionine/homocysteine ratio, which, as we saw in Chapter 7, is a fairly reliable and accessible measure for assessing methylation capacity and the one I use with my clients. The results showed that this ratio decreased considerably after the activity compared to pre-exercise levels. Both a reduction in methionine (indirectly reflecting a drop in SAMe) and a significant increase in homocysteine were observed. This confirmed a high consumption of SAMe during intense physical activity, evidenced both by its post-exercise reduction and by the rise in homocysteine, which, as we recall, is a byproduct of SAMe's action as a methyl group donor in any methylation process.

It is important to clarify that this decline in global methylation capacity after exercise is very short-term, occurring immediately after the activity ends. However, just like the increases in inflammation, lactate, and oxidative stress, this stimulus produced by exercise, over the medium and long term, results in the opposite: an increase in methylation capacity, a reduction in systemic inflammation, lower baseline levels of lactate, and an enhancement of antioxidant capacity.

Let's move forward. Now, let's discuss the first of the two most valuable findings from this study. What happens to the homocysteine generated from methylation? Of all the possible pathways the body could take, which one does it prioritize? The transsulfuration pathway (see Fig. 3b) is favored, synthesizing cysteine, which is then used to produce the antioxidants glutathione and taurine, whose levels increased by 17% and 85%, respectively.

---

[308] https://doi.org/10.3389/fphys.2020.609335

The second significant finding of the study (attention, athletes seeking to improve performance) was that the group of athletes who reversed the decline in methylation capacity within two hours after intense physical activity showed, in an endurance test the following day, an 8.5% improvement in performance compared to the group that did not manage to reverse the decline. Moreover, this group also showed a greater increase in cysteine levels (the precursor to antioxidants). In other words, they not only improved their performance and recovered faster, but also generated more antioxidants.

PUTTING IT ALL IN CONTEXT

1. Why is such a large amount of SAMe consumed during intense physical activity?

Although we do not yet have direct scientific evidence to answer this question, using common sense and drawing on what we learned about the various functions of SAMe in Chapters 2 and 3, my hypothesis is that the body primarily uses SAMe for **creatine synthesis**.

Creatine is a key molecule for the storage and rapid supply of energy in muscle cells, especially during high-intensity exercise. During intense physical activity, the energy demand in muscle cells increases significantly, and creatine plays a crucial role in the rapid regeneration of adenosine triphosphate (ATP) through the phosphocreatine system. Stored as phosphocreatine, it can donate its phosphate group and quickly convert adenosine diphosphate (ADP) into ATP, the body's main source of cellular energy.

In summary, intense exercise increases the demand for ATP, which, in turn, consumes creatine. To maintain adequate levels of creatine in the muscles, the need for its synthesis rises, consuming a large amount of SAMe and reducing its availability after exercise.

There are other possible, non-mutually exclusive explanations:

- **Phosphatidylcholine synthesis**: During intense exercise, muscle cells experience significant stress, and phosphatidylcholine helps maintain membrane integrity and neutralize free radicals, preventing cellular damage, reducing inflammation, and accelerat-

ing recovery. It also facilitates the mobilization and utilization of fat as fuel, enhancing metabolic flexibility. Phosphatidylcholine is a precursor to the neurotransmitter acetylcholine, which is crucial for muscle contraction and neuromuscular coordination. All of this increases the demand for phosphatidylcholine, which also requires SAMe for its synthesis.
- **Conversion of norepinephrine into adrenaline**: During intense physical activity, the sympathetic nervous system activates, increasing the release of catecholamines (norepinephrine and adrenaline). The conversion of norepinephrine into adrenaline, which requires SAMe, is essential for the "fight or flight" response that prepares the body for extreme physical efforts by raising heart rate, blood pressure, and glucose availability.
- **Repair of muscle tissue**: Intense exercise causes microtears in muscle fibers, requiring increased protein synthesis for muscle repair and growth. The methylation of RNA and proteins is crucial for the synthesis and regulation of new muscle proteins, processes that also consume SAMe.

2. Why does the body prioritize the synthesis of antioxidants over the synthesis of SAMe?

All of the previously mentioned processes that consume SAMe generate homocysteine as a byproduct. We know that homocysteine can follow two pathways to avoid accumulation and prevent toxicity:

1. the remethylation pathways (see Fig. 3a), which transform it into methionine for the synthesis of more SAMe, and
2. the transsulfuration pathway (see Fig. 3b), which converts it into cysteine for the synthesis of the antioxidants glutathione and taurine.

The interesting question now is: given that exercise depletes SAMe, why, once homocysteine has been generated, does the body not prioritize the synthesis of more SAMe and instead favor the transsulfuration pathway to synthesize the antioxidants glutathione and taurine?

Although we do not know for certain, it is reasonable to assume that it is due to the immediate need to neutralize oxidative stress—in other words, to minimize the impact of the increased production of free radicals (reactive oxygen species [ROS]). We know that the increase in ROS during exercise can cause cellular damage if not adequately neutralized, and for that, glutathione is essential, as it is the primary intracellular antioxidant.

But what about taurine, which increases even more than glutathione?

Taurine performs its functions at the extracellular level:

- **In muscle**, it not only protects against oxidative stress but also enhances contractile function.
- **In the brain (central nervous system)**, it acts as a neurotransmitter and neuromodulator in nerve signal transmission and the regulation of neuronal activity.
- **In the heart and blood vessels (cardiovascular system)**, it helps regulate myocardial contractility, blood pressure, and the prevention of arrhythmias.

From all of the above, the most important takeaway is to remember two key points:

1. During intense physical exertion, the body consumes a great deal of SAMe to initiate methylation processes, primarily to synthesize creatine, adrenaline, phosphatidylcholine, and other molecules. Later, with the homocysteine resulting from these processes, it synthesizes antioxidants to counteract the short-term oxidative stress caused by physical activity and to enhance muscle contractility, among other effects.
2. The more we can support the body in these tasks (by increasing methylation capacity during exercise and minimizing its loss afterward), the greater the improvement in physical performance, thanks to faster energy recovery, better muscle repair, and a stronger antioxidant effect.

Now, let's explore what we can do and what tools we have available to achieve this.

HOW TO INCREASE METHYLATION CAPACITY TO IMPROVE PERFORMANCE AND SPEED UP RECOVERY

Creatine

During intense physical activity, we deplete our levels of SAMe, and if this molecule is primarily used to synthesize creatine, the first option to consider is creatine supplementation. And yes, it is always my first option.

The benefits of creatine in strength sports and short-duration explosive activities are well documented and supported by numerous studies, so I will not dwell on that here. The doubt that once existed was whether creatine also provided benefits in longer-duration physical activities. Until a few years ago, my intuition was that it did, especially given its positive effects on systemic health outside of a sports context, but I hesitated to give a definitive answer because the scientific evidence was not yet abundant or solid enough to recommend it for sports with a more aerobic than anaerobic component. There was also the issue of weight gain. Creatine increases the water content within cells, and this water retention, once stores are saturated, could lead to a weight increase of about 1 to 2 kilograms (2.2 to 4.4 pounds).

However, I now have a clearer opinion and am convinced that creatine also provides benefits in longer-duration physical activities. My view was shaped by the study from the University of Oslo that we discussed earlier. Why? Because the tests in that study lasted at least one hour, and if the body tends to synthesize creatine during that time, it seems logical to think that it is because it needs it.

We have already discussed the possible reasons. To begin with, creatine increases energy availability (ATP), but it also offers indirect benefits. It improves muscle recovery by reducing muscle damage and inflammation between training sessions, allowing for more intense and frequent workouts. Moreover, it is also utilized by the brain, enhancing cognitive function, especially under stress or fatigue. It is well known that mental fatigue affects performance, and creatine possesses antioxidant properties that can help reduce oxidative damage in brain cells. In situations of high pressure and fatigue, creatine can help maintain mental clarity and the ability to make quick and accurate decisions—an essential advantage in competitive sports.

Regarding the issue of weight gain due to greater water retention within muscle cells, I believe that this potential drawback is offset by the osmotic effect, which improves hydration. Enhanced intracellular hydration can help maintain muscle function and prevent dehydration, which is especially beneficial during long-duration events. Additionally, creatine also improves thermoregulation, which is crucial in prolonged competitions.

Furthermore, I also believe that the clear benefits of creatine in very high-intensity, short-duration activities can extend to longer-duration activities.

First, in sports like cycling or trail running, it is common to alternate periods of moderate effort with anaerobic bursts of high intensity, such as during a sprint or an uphill climb.[309]

Second, even in competitions where a relatively steady pace is maintained, such as in a marathon, athletes pushing to their limits at the end of the race reach a point of exhaustion, and the body attempts to draw energy from every available source: glucose, fat, lactate, and —if available—phosphocreatine.

Regarding the timing of supplementation, we have two options:

1. supplement chronically with maintenance doses to keep creatine stores consistently saturated, or
2. supplement acutely with a single dose for a specific activity.

Given that creatine offers additional health benefits beyond sports performance—especially in the context of epigenetics by sparing SAMe, as we have explained—I prefer the first option, which is the one I use for myself and my athletes.

---

[309] https://doi.org/10.1080/15502783.2023.2204071

However, it has been shown that an acute single dose can also provide significant benefits for a specific moment, as demonstrated by a recent 2024 study[310] conducted by the Institute of Neuroscience and Brain Molecular Organization in Germany. A group of individuals, after 21 hours of sleep deprivation, was divided into two groups: one received a single high dose of creatine (0.35 g/kg of body weight —slightly more than four times the usual daily maintenance dose), while the other group received a placebo. The group that took creatine showed a clear improvement in cognitive performance and processing speed and reversed the metabolic alterations and cognitive decline associated with fatigue. The peak effect of this high dose occurred around four hours after ingestion, which should be considered when planning the timing of acute creatine supplementation.

If we are already supplementing daily, we can take the daily dose one hour before starting the intense activity, since being a smaller dose (about four times lower than the acute dose), it reaches its maximum absorption within one hour. After the intense activity, we can take the usual daily dose. In other words, if we supplement daily, on days of high-intensity physical activity, we can double the usual daily maintenance dose: one dose before exercise, to support methylation processes and benefit from its effects during the activity, and another dose afterward, to reduce the loss of methylation capacity and speed up recovery.

Phosphatidylcholine

Since the synthesis of phosphatidylcholine from phosphatidylethanolamine also consumes SAMe, a second option for conserving SAMe (and phosphatidylethanolamine) and increasing methylation capacity would be to provide the body with phosphatidylcholine directly.

A study[311] from the University of Colorado showed that phosphatidylcholine and phosphatidylethanolamine levels in muscle were higher in endurance-trained athletes compared to individuals with obesity and type 2 diabetes, and that this conferred better insulin

---

[310] https://doi.org/10.1038/s41598-024-54249-9
[311] https://doi.org/10.1152/japplphysiol.00664.2015

sensitivity. This improvement in insulin sensitivity has a clear physiological basis. Both phospholipids contribute to maintaining cellular membrane fluidity and stability, which facilitates the functionality of receptors, such as insulin receptors, thereby improving hormonal signaling and increasing glucose uptake. Additionally, phosphatidylethanolamine is involved in mitochondrial biogenesis and efficiency, which is crucial for efficient energy metabolism and insulin sensitivity.

However, immediately after exercise (90 minutes at 50% of VO2 max) in endurance-trained athletes—but not in the group of obese and diabetic individuals—a decrease in both phospholipids is observed, which persists for at least two hours after completing the activity. This decrease in athletes, compared to those with obesity or diabetes, is due to their greater efficiency in repairing and maintaining cellular membrane integrity, which involves a higher utilization of these phospholipids. However, it is important to highlight that this is only an acute and temporary response to exercise, since, in the long term, phospholipid levels increase due to the rise in the number of mitochondria in skeletal muscle as an adaptation to chronic exercise. Both phosphatidylcholine and phosphatidylethanolamine are essential components of mitochondrial membranes. In other words, the long-term increase in these phospholipids detected in athletes reflects a greater need to form new mitochondrial membranes as part of cellular adaptations to physical training.

A study from the University of Oslo detected the same decrease in both phospholipids during another test, where the intensity was increased to 75% of VO2 max, participants ingested carbohydrates before the activity, and the exercise duration was reduced to 45 minutes.[312] The difference in this case was that the most pronounced decline occurred in phosphatidylcholine. This also makes sense, as increasing intensity and shifting the energy substrate (from lower fat consumption to greater glucose use) generate more oxidative damage, and the body consumes phosphatidylcholine (and SAMe) to minimize damage to cellular membranes.

[312] https://doi.org/10.1038/s41598-018-24976-x

On the other hand, if we do not use the enzyme phosphatidylethanolamine N-methyltransferase (PEMT) to create phosphatidylcholine (see Fig. 2) because we are already providing it exogenously—and thus saving phosphatidylethanolamine—we will also enhance mitochondrial respiration and ATP levels while reducing glucose production through gluconeogenesis, which in turn increases insulin sensitivity.[313]

A reasonable daily dose of phosphatidylcholine could range between 1.5 and 3 grams. Since soy lecithin contains approximately 25% phosphatidylcholine and 20% phosphatidylethanolamine, a good option is to consume between 6 and 12 grams of soy lecithin per day, as I explained in Chapter 8, or a single dose of about 5 grams around five hours before a competition or high-intensity training session.

As with creatine, I prefer the option of a daily maintenance dose rather than an acute pre-exercise dose. Moreover, those daily grams of lecithin also provide about 750 milligrams of another very interesting phospholipid: phosphatidylserine.

A study[314] from the University of Wales demonstrated that supplementation with that amount of phosphatidylserine significantly increased exercise time to exhaustion at an intensity of 85% of VO2 max (approximately the second threshold). Most previous studies on phosphatidylserine have focused on its effects on cognitive functions, showing improvements in memory and mood under stress, but this study suggests that it could also have an ergogenic effect. This is because phosphatidylserine activates enzymes such as adenosine triphosphatase (ATPase), which enhances calcium reuptake in muscle cells and maintains a proper balance of ions (calcium, sodium, and potassium). This balance is crucial for muscle function, allowing for better contraction and relaxation of the muscles during exercise, which translates into a greater capacity to perform high-intensity exercise before reaching exhaustion.

---

[313] https://doi.org/10.2337/db13-0993
[314] https://doi.org/10.1249/01.mss.0000183195.10867.d0

SAMe

If the body consumes large amounts of SAMe during intense physical activity to the point of reducing methylation capacity, and if reversing this decline is beneficial, why is supplementing directly with SAMe not my first option?

First, because the body primarily uses SAMe to synthesize creatine and phosphatidylcholine (lecithin), and we can provide both molecules directly.

Second, I am generally reluctant to use supplements containing molecules that do not naturally occur in foods, since the body is not accustomed to ingesting them and, in addition, there are often no long-term studies confirming their safety. As I mentioned in Chapter 8, although SAMe is available over the counter, I almost consider it a drug and believe it should be used with caution.[315]

Personally, I use it only at the lowest doses, and only when the potential benefit clearly outweighs the possible risk, and in people who have been unable to increase their methylation capacity with SAMe-sparing molecules that are naturally present in foods, such as creatine and phosphatidylcholine.

Nevertheless, since we've seen that the body likely consumes SAMe for more than just the synthesis of creatine and phosphatidylcholine, I see no issue with supporting it with a low dose (200 mg) of SAMe occasionally, before a particularly demanding or important competition or training session. Peak plasma concentrations of SAMe are reached about five hours after a fasting dose,[316] so it is important to plan its intake according to the timing of the competition or training in order to maximize its benefits.

Protein

I do not recommend protein supplementation to enhance performance before or during high-intensity training or competition. However, it is essential to ensure an adequate daily intake of high-quality protein through food. This is fundamental for muscle synthe-

---

[315] https://doi.org/10.1038/s42003-022-03280-5#Sec16
[316] https://doi.org/10.1186/s40360-020-00466-7

sis and for repairing the microtears in muscle fibers, which accelerates recovery.

On a regular basis, I usually recommend a protein intake in the range of 1.2 to 1.6 g/kg of body weight (approximately 0.54 to 0.73 g/lb). However, on days when training has been more intense than usual, I advise ensuring an intake between 1.6 and 1.8 g/kg (roughly 0.73 to 0.82 g/lb) of body weight.

In ultradistance events, the situation changes due to the prolonged duration of physical activity. In these cases, I do recommend supplementing with whey protein during the event to prevent muscle catabolism and to support SAMe synthesis and methylation capacity, thanks to the methionine content in whey.

HOW TO SUPPORT ANTIOXIDANT SYNTHESIS

Everything we've discussed so far has focused on enhancing the benefits of reversing the loss of methylation capacity (SAMe) due to the increased demand that occurs during and after exercise. However, we have also seen that, once methylation has occurred and the inevitable homocysteine has been generated, the body prioritizes the pathway that converts it into antioxidants (see Fig. 3b).

If we support the body in this task, we gain a twofold benefit: First, because we are aiding a process the body already considers a priority, and second, because by providing these antioxidants, the body can allocate more resources to other pathways—such as the remethylation pathway (see Fig. 3a)—to generate more SAMe and indirectly reinforce its methylation capacity.

Vitamin B6

As we saw in Chapter 10, for homocysteine to be converted into cysteine and then into the antioxidants glutathione and taurine, the first thing we must ensure is that there is enough vitamin B6 available. Supplementation is not necessary unless it is to maintain blood levels of B6 within its optimal range (50-170 nmol/L).

Now, let's move on to analyze the possible molecules that could assist the body's need to synthesize antioxidants.

N-acetylcysteine

Since cysteine is the precursor of both glutathione and taurine (see Fig. 3b), the first step in supporting antioxidant synthesis would be to consider increasing cysteine levels. However, direct cysteine supplementation is not efficient, so we have a better option: N-acetylcysteine (NAC), which has greater bioavailability (is better absorbed) and is more stable (less susceptible to oxidation) than cysteine.

As a supplement, NAC is effective at raising glutathione levels. It is also used as a mucolytic drug, as it breaks the strong bonds in the proteins of thick, sticky mucus, making it more liquid and easier to expel from the lungs through coughing.

Regarding exercise performance, research on NAC is limited, but there is enough evidence to draw some conclusions.

A double-blind study[317] conducted with nine well-trained cyclists, who performed 6 sets of 5 minutes at high intensity followed by a 10-minute time trial, showed several effects of NAC compared to a placebo:

- **Decreases lipid peroxidation**: NAC reduced oxidative damage to fats in cellular membranes, which is not surprising given its effect on raising the antioxidant glutathione.
- **Increases fat oxidation and reduces lactate**: NAC altered metabolism, favoring the use of fat as the primary energy source and sparing glucose, keeping it in the blood.
- **Reduces median electromyography (EMG) frequency in the quadriceps**: EMG measures the electrical activity produced by muscles during contraction. The reduction in median frequency reflects a shift toward using slow muscle fibers rather than fast ones, improving fatigue resistance but reducing maximal power and explosive performance.
- **Decreases average power output during the 10-minute time trial**: Since high-intensity exercise requires the use of glucose and lactate, as well as fast-twitch muscle fibers, this was expected, considering the previous three points.

---

[317] https://doi.org/10.1139/apnm-2012-0482

This set of data suggests that NAC might not be useful for short-duration physical activities with an anaerobic component but could be interesting for long-distance events, where the use of slow-twitch fibers, fat as a fuel source, and the conservation of glucose and glycogen are crucial.

Another study[318] conducted with well-trained cyclists, involving about 50 minutes of activity (45 minutes at an intensity close to the first threshold and 5-6 minutes at near-maximum intensity until exhaustion), showed that prior administration of NAC prolonged time to fatigue by 23.8% compared to placebo. NAC also slowed the decrease in sodium-potassium pump activity and the rise in plasma potassium levels. Let's explain this in more detail.

The sodium-potassium pump is a vital protein in cells that regulates the balance of sodium and potassium by moving sodium out of the cell and potassium into the cell. This process is essential for many cellular functions, including the transmission of nerve impulses and muscle contraction, as it helps maintain proper concentrations of these ions inside and outside the cells.

During intense or prolonged exercise, muscle cells work hard and consume a lot of energy, which leads to an increase in the generation of waste products, including ROS (reactive oxygen species). These ROS can damage the sodium-potassium pump, reducing its efficiency and contributing to muscle fatigue. When the activity of this pump declines, sodium accumulates inside the cell and potassium accumulates outside the cell, negatively affecting muscle performance.

Excess sodium inside the cell can cause cellular swelling, disrupt the structure and function of the cell, and impair the generation and transmission of the electrical impulses necessary for muscle contraction.

On the other hand, if extracellular potassium (in plasma) increases, the ability of muscle cells to generate and transmit electrical signals decreases, resulting in weaker contractions and greater fatigue during exercise.

In summary, this study shows that ROS (reactive oxygen species) play an important role in muscle fatigue by reducing the efficiency of

---

[318] https://doi.org/10.1113/jphysiol.2006.115352

the sodium-potassium pump and that antioxidants, such as NAC (a precursor of glutathione and taurine), help minimize this impact.

However, it is important to point out that in this study, NAC administration was acute, given one hour before physical activity, at an extremely high dose (about nine times higher than the usual daily dose) and delivered intravenously. In everyday practice, this is not feasible, so it might make more sense to supplement with normal doses of NAC on a regular basis.

The study mentioned was conducted during an activity of moderate duration (about 50 minutes), where 90% of the time was spent below the first threshold. But what happens in much shorter, more explosive events? As observed in the first study, NAC favored fat usage over glucose and a greater reliance on slow-twitch fibers—something undesirable in highly anaerobic events.

Another study[319] found that after higher-intensity, shorter-duration exercises—such as 20 repetitions of 20 seconds with 2 minutes of rest—although NAC reduced lactate levels and hydrogen ion accumulation, it did not improve performance. Or perhaps it was precisely because of that.

This study also provides additional evidence that lactate is not the cause of muscle fatigue—at least during high-intensity exercise—as was once commonly believed. Instead, the accumulation of potassium in the plasma appears to play a more significant role. Extracellular potassium makes it more difficult for the membrane surrounding each cell (the sarcolemma) to generate the electrical signals needed for muscle cells to contract with the same strength or frequency as they normally would. This leads to a quicker onset of fatigue during exercise.

A 2023 meta-analysis[320] confirms that NAC supplementation can be beneficial in endurance sports, where oxidative damage—particularly to the sodium-potassium pump—can negatively impact performance. However, in explosive and short-duration activities, where oxidative fatigue is not the primary limiting factor, NAC administration might not be beneficial.

---

[319] https://doi.org/10.3390/antiox12010053
[320] https://doi.org/10.3390/nu15112463

NAC supplementation is typically done in capsules or powder, at a daily dose between 500-1500 mg. Although its safety profile has been confirmed for up to one year at these doses, as with SAMe, NAC is not a molecule naturally present in foods, meaning humans have not evolved to ingest it naturally. For this reason, it would not be my first choice among the available strategies. When aiming to enhance antioxidant capacity, unless there are more long-term studies or a pathology where its potential benefit is very high, I prefer to opt for other strategies, which we will continue to explore and summarize at the end.

Glutathione

If reducing oxidative damage by boosting antioxidant capacity improves performance—and if the potential benefit of NAC lies in its ability to serve as a precursor to the antioxidants glutathione and taurine—it is crucial to consider supplementing with the latter. Let's start with glutathione.

Although there are companies that market glutathione, unlike NAC, which I have included as a possible option, glutathione supplementation is something I clearly do not recommend.

There are three main reasons for this.

The first is that enzymes in the intestine break down glutathione, which significantly reduces the amount of intact glutathione that can be absorbed and used by the body. For this reason alone, glutathione supplementation makes little sense—especially when there are other options like NAC, which, as a precursor molecule, has been proven to increase endogenous glutathione synthesis. Since it is synthesized by the body itself, it does not require absorption, ensuring much greater effectiveness.

The second reason is that glutathione is not a molecule found in foods. Although this fact alone is not a definitive reason to avoid a supplement, it is a strong argument for exercising extreme caution before using it. The body is not evolutionarily adapted to receiving glutathione exogenously, so the benefit-to-risk ratio must be very high to justify its use.

The third reason is that supplemented antioxidants—including glutathione—[321] have been shown to be ineffective and may even interfere with the beneficial effects of exercise.[322] This makes perfect sense from a physiological standpoint. It's nearly impossible to get the dosage exactly right; supplementation can neutralize too many free radicals, which can be harmful, since these also have beneficial roles in small amounts, such as signaling key cellular processes and activating protective immune mechanisms. In other words, supplemented antioxidants can disrupt the body's natural balance. In contrast, endogenously synthesized antioxidants are more effective and safer because they are precisely matched to the body's physiological needs in both quantity and location, ensuring effectiveness without adverse effects.

Therefore, the ideal approach is to provide the body with the necessary tools and components (precursors) so it can synthesize exactly the amount of glutathione it determines is needed. We've already seen one way to increase one of these precursors: cysteine—either by increasing SAMe, enhancing methylation capacity, and converting homocysteine into cysteine, or through NAC supplementation. Another important precursor is glycine, which I won't expand on here because it has already been covered in this book, as well as on my blog and in my previous works.

However, there is another strategy we can add: taurine supplementation. Part of the cysteine available to the body is used to synthesize taurine (see Fig. 3b). By supplementing with taurine, we relieve the body of this function, saving cysteine and thereby indirectly increasing its availability for glutathione synthesis.

Let's now turn to taurine.

Taurine

Taurine, like glycine, is an amino acid and one of my favorite supplements. It occurs naturally in food, and the body is evolutionarily adapted to ingest it. As we've already mentioned, one of its main roles is to act as an extracellular antioxidant, protecting cells from oxidative

---

[321] https://doi.org/10.1089/acm.2010.0716
[322] https://doi.org/10.1073/pnas.0903485106

damage and helping maintain the integrity of cell membranes. However, taurine also plays many other essential roles in the body:[323]

- **Development and function of the central nervous system**: Taurine is essential for brain development, stabilization of neuronal cell membranes, and regulation of neurotransmitter receptor activity, as it helps balance neuronal excitation and inhibition.
- **Regulation of electrolyte balance**: It helps regulate calcium, potassium, and sodium levels in cells, which is critical for proper cellular function and dehydration prevention.
- **Cardiovascular functions**: Taurine has protective effects on the heart and blood vessels, improving myocardial contractility, stabilizing heart rhythm, and helping to regulate blood pressure.
- **Fat metabolism**: It plays a role in fat digestion as part of bile acids, which are essential for emulsifying and absorbing lipids in the intestine. It also helps regulate cholesterol levels.
- **Immunomodulation**: Taurine supports immune system regulation by enhancing white blood cell function and reducing inflammation.
- **Eye health**: This amino acid is crucial for retinal protection, as it is abundant in the retina and helps shield retinal cells from oxidative damage.

Because of these important functions—and given that declining taurine levels over time are associated with age-related diseases—it is now being actively studied as a promising anti-aging molecule.[324] A 2023 study[325] conducted on monkeys and mice showed that taurine supplementation, aimed at reversing age-related decline, increased longevity and improved health through various mechanisms: it reduced cellular senescence, protected against telomerase deficiency, suppressed mitochondrial dysfunction, decreased DNA damage, and attenuated age-related inflammation.

---

[323] https://doi.org/10.3390/nu15194236
[324] https://doi.org/10.3390/nu15194236
[325] https://doi.org/10.3390/nu15194236

Taurine is also being investigated as a potential tissue protector against metal toxicity[326] and as a treatment for post-COVID, since raising taurine levels has been shown to alleviate symptoms and reduce adverse clinical events.[327] The same has been observed for various types of infections.[328]

In the context of exercise, we've already mentioned that taurine synthesis becomes a priority for the body during intense physical activity, which gives us a clear sense of its importance—as highlighted in this 2021 systematic review.[329]

Let's take a look at some of its specific benefits:

- **Improves muscle contraction**: Taurine helps regulate calcium in muscle cells, which is essential for effective muscle contraction.
- **Reduces muscle fatigue**: Its antioxidant properties help lower oxidative stress and cellular damage, delaying the onset of fatigue.
- **Accelerates muscle recovery**: Taurine reduces inflammation and oxidative damage in muscle cells, lowering markers of muscle damage such as creatine kinase (CK) and lactate dehydrogenase (LDH). This leads to faster recovery and a reduction in delayed onset muscle soreness (DOMS).
- **Increases aerobic capacity**: It enhances the body's oxygen utilization efficiency by optimizing mitochondrial function, and helps maintain a stable heart rate and efficient cardiac function.
- **Improves energy metabolism**: Taurine supports cellular energy production, facilitates nutrient transport, and increases glucose uptake by muscle cells.

Most of the available studies have investigated taurine in the context of enhancing aerobic capacity. However, more recent research

---

[326] https://doi.org/10.1007/s12011-024-04191-8
[327] https://doi.org/10.1371/journal.pone.0304522
[328] https://doi.org/10.1007/s44254-024-00055-5#Sec11
[329] https://doi.org/10.3389/fphys.2021.700352#h6

suggests there may be some improvement in anaerobic power at maximal intensities when taurine is taken as an acute dose before exercise.[330] Additionally, taurine works synergistically with caffeine, proving more effective than either alone or placebo.[331]

That said, I do not recommend its use for very high-intensity activities with a predominantly anaerobic component. The reason is that taurine lowers heart rate by improving cardiovascular efficiency, meaning the heart works more effectively to pump blood and oxygen to the muscles. This is beneficial for endurance and aerobic capacity in longer-duration efforts (submaximal intensities), allowing sustained performance with less premature fatigue. However, this same effect could limit the ability to reach maximum heart rate, which may be detrimental in short-duration, purely anaerobic events.

In hot conditions, taurine has been shown to increase time to exhaustion and reduce the rate of perceived exertion (RPE), while also lowering core body temperature.[332]

When it comes to supplementation, the daily taurine dose should aim to maintain optimal serum levels, which range from 75-140 µmol/L (or nmol/mL). In most people, this can be achieved with an intake of 3-6 g per day (80 mg/kg of body weight), divided into three separate doses throughout the day. However, before supplementing, I always consider it essential to check a person's current levels through bloodwork to ensure they are not already in the high-optimal range —or even above it—in which case supplementation would not be advisable.

If opting for an acute dose prior to high-intensity exercise, it should be half of the daily dose—about 40 mg/kg. If following this strategy and already supplementing daily, that amount should be subtracted from the usual daily intake for that day. Taurine reaches its peak absorption approximately 60 to 90 minutes after ingestion. If combining it with caffeine, keep in mind that caffeine's peak plasma levels occur 45 to 60 minutes after ingestion—this is when it crosses the

---

[330] https://doi.org/10.5114/biolsport.2023.119990
[331] https://doi.org/10.3390/nu14204399
[332] https://doi.org/10.1080/17461391.2019.1578417

blood-brain barrier and begins binding to adenosine receptors in the brain. A good strategy would be to take both substances one hour before starting the warm-up.

In food, taurine is found mainly in oysters, animal heart and liver, as well as in octopus and venison. These were common staples in our ancestors' diets but are no longer among the most frequently consumed foods for most people today.

SUMMARY AND CONCLUSIONS

In my view, when we talk about nutrition in the context of intense physical activity, the most important thing for minimizing fatigue and maximizing performance is to replenish what we lose during exercise. So, what do we lose? There are six key elements. Four of them are already fairly well understood and were covered in depth in my blog and in my previous book, *Evolutionary Nutrition for Athletes*:

1. water
2. sodium
3. glucose/glycogen
4. fat

Of these four, fat is the least concerning during physical activity, since we have enough stored in the body to avoid running out.

The two types of loss that were not as well understood are the ones I've focused on extensively in this book, and more specifically in this appendix:

1. methylation capacity (including SAMe, creatine, and phospholipids).
2. antioxidant capacity (specifically glutathione and taurine).

To minimize this impact, the first step is to support the body's methylation process—especially by supplying the molecules that consume the most SAMe in their synthesis: creatine and phosphatidylcholine. As a byproduct of this methylation, homocysteine is generated, which is the precursor to cysteine, and in turn, cysteine is the precursor to the antioxidants glutathione and taurine (see Fig. 3b).

For homocysteine to be converted into cysteine, it is crucial to maintain optimal levels of vitamin B6. And if we supplement with taurine, we gain two benefits at once: on the one hand, we replenish the taurine lost during exercise, and on the other, we spare cysteine —since it no longer needs to be split between synthesizing both taurine and glutathione, more of it becomes available for glutathione synthesis.

REFERENCES

Dricu, A. (Ed.). (2012). *Methylation: From DNA, RNA and Histones To Diseases and Treatment.* InTech. https://doi.org/10.5772/2932

Dumitrescu, R. G. & Verma, M. (Eds.). (2018). *Cancer Epigenetics for Precision Medicine: Methods and Protocols* (Methods in Molecular Biology, vol. 1856). Humana Press. https://doi.org/10.1007/978-1-4939-8751-1

Ferguson, L. R. (2014). *Nutrigenomics and Nutrigenetics in Functional Foods and Personalized Nutrition.* CRC Press.

Kohlmeier, M. (2013). *Nutrigenetics: Applying the Science of Personal Nutrition.* Academic Press. https://doi.org/10.1016/C2010-0-67181-1

López-Otín, C. (2019). *La vida en cuatro letras: Claves para entender la diversidad, la enfermedad y la felicidad.* Ediciones Paidós.

__. (2021). *Egoístas, inmortales y viajeras.* Ediciones Paidós.

Neidhart, M. (2016). *DNA Methylation and Complex Human Disease.* Academic Press. https://doi.org/10.1016/C2013-0-13028-0

Niculescu, M. D. & Haggarty, P. (Eds.). (2011). *Nutrition in Epigenetics.* Wiley-Blackwell.

Patel, V. B. & Preedy, V. R. (Eds.). (2019). *Handbook of Nutrition, Diet, and Epigenetics.* Springer.

Peedicayil, J., Grayson, D. R. & Avramopoulos, D. (Eds.). (2014). *Epigenetics in Psychiatry.* Academic Press. https://doi.org/10.1016/C2020-0-00389-0

Sahebkar, A. & Sathyapalan, T. (Eds.). (2021). *Natural Products and Human Diseases: Pharmacology, Molecular Targets, and Therapeutic Benefits* (Advances in Experimental Medicine and Biology, vol. 1328). Springer. https://doi.org/10.1007/978-3-030-73234-9

Simopoulos, A. P. & Milner, J. A. (Eds.). (2010). *Personalized Nutrition: Translating Nutrigenetic / Nutrigenomic Research into Dietary Guidelines* (World Review of Nutrition and Dietetics, vol. 101).

Karger. https://doi.org/10.1159/isbn.978-3-8055-9428-8

WEST, B. J. (Ed.). (2013). *Fractal Physiology and Chaos in Medicine* (2.ª ed.) (Studies of Nonlinear Phenomena in Life Science, vol. 16). World Scientific.

YASUI, D. H., PEEDICAYIL, J. & GRAYSON, D. R. (Eds.). (2017). *Neuropsychiatric Disorders and Epigenetics* (Translational Epigenetics Series). Academic Press. https://doi.org/10.1016/C2013-0-15470-0

YU, B. P. (Ed.). (2015). *Nutrition, Exercise and Epigenetics: Ageing Interventions* (Healthy Ageing and Longevity, vol. 2). Springer. https://doi.org/10.1007/978-3-319-14830-4

LIST OF FIGURES

FIGURE 1	Polyamine Synthesis	56
FIGURE 2	Methionine Cycle (Methylation Process)	66
FIGURE 3A	Homocysteine Cycle (Remethylation Pathways)	67
FIGURE 3B	Homocysteine Cycle (Transsulfuration Pathway)	68
FIGURE 4	Pyramid of Life	132

ABBREVIATIONS, ACRONYMS, AND INITIALISMS

5-MTHF	5-methyltetrahydrofolate
5mC	5-methylcytosine
ADHD	Attention-deficit/hyperactivity disorder
ADMA	Asymmetric dimethylarginine
ADP	Adenosine diphosphate
AGAT	L-arginine:glycine amidinotransferase
AHCY	Adenosylhomocysteinase
α-KG	Alpha-ketoglutarate
ALS	Amyotrophic lateral sclerosis
AML	Acute myeloid leukemia
AMPK	AMP-activated protein kinase
Apo-E	Apolipoprotein E
APP	Amyloid precursor protein
ASD	Autism spectrum disorder
ATP	Adenosine triphosphate
ATPase	Adenosinetriphosphatase
BCM-7	Bioactive molecule beta-casomorphin-7
BDNF	Brain-derived neurotrophic factor
BHMT	Betaine-homocysteine methyltransferase pathway
BPA	Bisphenol A
CaAKG	Calcium alpha-ketoglutarate
CBS	Cystathionine beta-synthase
Cdk5	Cyclin-dependent kinase 5
CK	Creatine kinase
CNS	Central nervous system
COMT	Catechol-O-methyltransferase
CVD	Cardiovascular diseases
DAO	Diamine oxidase

dcSAMe	Decarboxylated SAMe
DHA	Docosahexaenoic acid
DNMT	DNA methyltransferase
DOMS	Delayed-onset muscle soreness
EDHF	Endothelium-derived hyperpolarizing factor
EGCG	Epigallocatechin gallate
EMG	Electromyography
ER	Estrogen receptors
ERα+	Estrogen receptor alpha-positive
EZH2	Histone-lysine methyltransferase
g	Grams
GABA	Gamma-aminobutyric acid
GAMT	Guanidinoacetate N-methyltransferase
GLUT1	Brain glucose transporter at the blood-brain barrier
GlyT1	Glycine transporter 1
GMP-7	Gliadorphin-7
GNMT	Glycine N-methyltransferase
GPx	Glutathione peroxidase
GSH	Reduced glutathione
GSR	Glutathione reductase
GSSG	Oxidized glutathione
H_2S	Hydrogen sulfide
HAT	Histone acetyltransferase
HDAC	Histone deacetylase
HDACi	Histone deacetylase inhibitors
HNMT	Histamine N-methyltransferase
hTERT	Human telomerase reverse transcriptase
HTT	Huntingtin
IgE	Immunoglobulin E
IP	Identity preserved
kg	Kilograms
L-DOPA	Levodopa
LAT1	L-type amino acid transporter 1
LDH	Lactate dehydrogenase
LDL	Low-density lipoprotein cholesterol
MAO	Monoamine oxidase

MAT	Methionine adenosyltransferase
MLL-R	Mixed lineage leukemia-rearranged
mRNA	Messenger RNA
MS	Multiple sclerosis
mTOR	Mammalian target of rapamycin
MTR	Methionine synthase (5-methyltetrahydrofolate-homocysteine methyltransferase), also known as cobalamin-dependent methionine synthase
NAC	N-acetylcysteine
NAD+	Nicotinamide adenine dinucleotide
NF-κB	Nuclear factor kappa-light-chain-enhancer of activated B cells
NMDA	N-methyl-D-aspartate receptors (a type of glutamate receptor)
NNK	Nicotine-derived nitrosamine ketone
NNK carcinogen	4'-(Methylnitrosamino)-1-(3-pyridyl)-1-butanone
NNMT	Nicotinamide N-methyltransferase
No GMO	Non-genetically modified organism
NO	Nitric oxide
Nrf2	Nuclear factor erythroid 2-related factor 2
OPC	Oligomeric proanthocyanidins (grape seed polyphenols)
P-5-P	Pyridoxal 5'-phosphate
PEMT	Phosphatidylethanolamine N-methyltransferase
PGC-1α	Peroxisome proliferator-activated receptor gamma coactivator 1-alpha
PNMT	Phenylethanolamine N-methyltransferase
PP2A	Protein phosphatase 2A
PRMT6	Protein arginine N-methyltransferase 6
PVC	Polyvinyl chloride
REM	Rapid eye movement (sleep phase)
ROS	Reactive oxygen species
RPE	Rate of perceived exertion
SAH	S-adenosylhomocysteine
SAMe	S-adenosylmethionine
SIBO	Small intestinal bacterial overgrowth
SLE	Systemic lupus erythematosus

TCR	T-cell receptors
Th2	T helper 2 cells
TMG	Trimethylglycine or betaine
TNF-α	Tumor necrosis factor alpha
VDR	Vitamin D receptor
VEGF	Vascular endothelial growth factor
VLDL	Very low-density lipoproteins
VO2 max	Maximal oxygen consumption
WHO	World Health Organization

Evolutionary Nutrition

Eat to Live, Not Just to Survive

Curro Clavero Valdivielso

Publication date: March 2025
192 pages
15,24 x 22,86 cm
ISBN: 978-84-09-69968-0

More than half of the population over the age of 15 has been diagnosed with at least one chronic disease: diabetes, cancer, cardiovascular disease, joint problems, respiratory conditions, autoimmune disorders, thyroid disease... dementia, migraines, depression...

This leads us to believe that having one of these diseases is normal and inevitable. But the truth is, just because something is common doesn't mean it's normal.

In my practice, I often hear things like: *"I don't think I eat poorly, and I do what I've always done, but my weight keeps going up,"* *"If I'm not at the doctor's for one thing, I'm there for another,"* *"I've been dragging this cold around for weeks,"* *"I have to take a pill for blood sugar, cholesterol, blood pressure, uric acid, sleep, pain...,"* *"I feel fatigued, drained, and constantly hungry."* But none of this is normal—and it is preventable.

Throughout this book, backed by the latest science and proven experience, you'll learn how to nourish yourself properly, giving your body what it truly needs to achieve optimal health. And you'll realize that doing so is much simpler than it seems.

Would you rather spend the rest of your years sick and on multiple medications, or full of energy with strong physical and mental faculties that enhance your quality of life and well-being?

Come on! Life is waiting for you

Evolutionary Nutrition for Athletes

Periodized Carbohydrate Nutrition: From Amateur to Champion

Curro Clavero Valdivielso

Publication date: April 2025
304 pages
15,24 x 22,86 cm
ISBN: 978-84-09-72301-8

E volutionary nutrition. Carbohydrate periodization. Two fuel tanks and two energy sources. Metabolic flexibility. The complete athlete. Variability. The adaptive organism. A healthy athlete… These are just some of the key topics that will be explored step by step throughout this book.

You'll see that physical training alone isn't enough to boost performance —what you eat and when you eat it play a crucial role in unlocking your full physical and mental potential.

You'll realize that eating the right foods at the right time not only enhances the quality of your training and your adaptations to it, but also improves your overall health—which, in turn, contributes to better performance. A double win.

You'll understand why the foods that have been with humankind the longest are the ones that suit us best.

You'll discover that, just as training should vary, so should your nutrition.

You'll learn how to make the most of your two main energy sources—fat and carbohydrates—so you can benefit from both worlds and become as efficient and well-rounded as possible.

This book brings you science—lots of it—a bit of theory, and, above all, practical tools. It's about translating cutting-edge research into everyday strategies, so you can fuel yourself to push further, perform better, and become the best version of yourself.

Love learning! Ready to level up? Let´s go!

Printed in Dunstable, United Kingdom